Bacteria In Control Of Life, Death, & Evolution?

A Dissertation By: Phyllis Abbott, Ph.D.

Bloomington, IN  Milton Keynes, UK
authorHOUSE®

AuthorHouse™
1663 Liberty Drive, Suite 200
Bloomington, IN 47403
www.authorhouse.com
Phone: 1-800-839-8640

AuthorHouse™ UK Ltd.
500 Avebury Boulevard
Central Milton Keynes, MK9 2BE
www.authorhouse.co.uk
Phone: 08001974150

First published by AuthorHouse 8/22/2006

ISBN: 1-4259-5336-0 (e)
ISBN: 1-4259-5330-1 (sc)

Library of Congress Control Number: 2006906736

Printed in the United States of America
Bloomington, Indiana

This book is printed on acid-free paper.

In loving memory of my Mother, Mary Jane O'Neill.

Once you carried me next to your heart.

Now I carry you next to mine.

What, you ask, was the beginning of it all?
And it is this............
Existence that multiplies itself
For sheer delight of being
And plunges into numberless trillions of forms
So that it might
find
itself
innumerably.

Sri Aurobindo

<u>Bacteria In Control Of Life, Death, & Evolution?</u>

by

Phyllis Abbott, Ph. D.

Abstract

Pathological bacteria are only 5% of the bacterial population. The other 95% promote the health and well-being of Earth. The digestive tract holds trillions of archaebacteria from over 4 1/2 billion years ago. When in danger, bacteria create shells for protection. Are humans evolved shells in order to protect the bacteria from atmospheric oxygen? Life forms are descended from prokaryote archaebacteria, for whom oxygen is unnecessary. After millions of years of evolution, can bacteria now direct humans to return the planet, through pollution, ozone depletion, or a nuclear disaster, to a more manageable level of oxygen from a present 21% to less than 1%? No bacteria reside in the cranial brain. Was the enteric nervous system the first brain? Are the archaebacteria within the gastrointestinal tract directing the actions of the body? Are the archaebacteria the architects and directors of evolution?

Chapter 1

Bacteria have long been associated with pathology but only about 5% of the bacterial population are harmful pathogens. The other 95% of bacteria form a natural and necessary part of the microorganisms that inhabit planet Earth. Bacteria are essential to all life forms, and even healthy humans could not exist without bacteria in, and on, their bodies. Bacteria are recognized as the first organisms to proliferate on earth, and all life forms are descended from them. Why are bacteria residents, or attached to, every life form? What role do they carry out that no organism can thrive or survive without them?

To counter practices of eliminating bacteria as only pathogenic, more emphasis should be placed on protecting and nurturing healthy bacteria. A proliferation of antibiotic medications, has led to the disturbances of bacterial communities within the human body. Each person's body has about 10 trillion human cells, but these are outnumbered by approximately 100 trillion microbial cells. Communities of microflora are established in and on many areas of the human body, and include the mouth and digestive tract. The overall health, development, and well being of each organism apparently depends on the active participation of every bacterial community within the organism

Dr. Lynn Margulis first proposed that life forms share a mutually beneficial relationship with bacteria. This relationship is named endosymbiosis, wherein each entity provides benefits to the other. The scientific community first rejected the endosymbiosis hypothesis but gradually, through years of scientific inquiry, rejection has changed to acceptance.

In the evolving and endosymbiotic relationship between bacteria and all life forms, the evolved form is considered to be the host. However, within each organism, bacteria reside in their original prokaryote form of over 4.5 billion years ago. It could be said that bacteria are in charge of life and death. They can create life by replication, and they can destroy life by secretions of toxins. Also, pathogenicity can develop through gene rearrangement and acquisition of DNA from other organisms during evolution (Arber, 2002).

Throughout the evolution of all the organisms that inhabit planet Earth, it appears that bacteria have maintained control of each organism in its methods of replication, in how its nutrients are obtained and absorbed, in how its respiratory and metabolic pathways are directed, all the while creating appropriate shelters so that each organism is protected and can survive in instances of adversity. In times of nutrient and energy shortages, bacteria become pathogenic by emitting toxins in order to reduce their population. However, the toxins can also be used to remove anything that threatens the survival of the bacteria.

Bacteria

Bacteria were the only form of life for 80% of earth's time-line. Although many life forms have evolved since then, the number of bacteria on our planet today "has been estimated at five to twenty-times the total mass of all animal life, both aquatic and terrestrial" (Postgate, 1992, p. 3). Bacteria appear to play an important role in every living organism from its inception, throughout its growth, and in its decay and death. No organism can grow and flourish without the presence of bacteria. The manipulation of bacteria is extremely important in medical fields, research laboratories, and in many areas of technology (Singleton, 1999).

Bacteria are the scientifically acknowledged prokaryote cells from which all life has evolved. Life requires energy to maintain and reproduce itself (Fortey, 1997). Bacteria are capable of obtaining energy from outside sources through either inorganic or organic compounds using anaerobic fermentation, aerobic respiration, or photosynthesis from sunlight (Oparin, 1953). Bacteria have a distinct advantage over all other organisms because, not only can some bacteria ingest organic compounds, others can break down inorganic materials (e.g. minerals) for their nutritional and energy needs (Postgate, 1992).

A bacterium, usually only a few micrometers (millionths of a meter) in size, has a limited ability to assimilate nutrients which have to be absorbed across the cell membrane. The smooth membrane around the cell only stretches to a certain volume (Davis & Solomon, 1986), but *Epulopiscium fischelsoni*, bacteria in sturgeon fish, measure 500 micrometers, because they have membranes that form folds. When the volume inside the bacteria of the sturgeon fish increases, the membranes stretch to accommodate the larger mass (Hrywana, 1996).

Each bacterium can reproduce and maintain itself, an ability known as autopoieisis (Margulis & Sagan, 1995). Evolved from inorganic elements, several kinds of bacteria transformed themselves to organic life. It is proposed that the change to organic life was accomplished originally through the actions of ribonucleic acid (RNA) and enzymes. RNA can produce its own proteins and enzymes, and, through enzymatic actions, RNA formed deoxyribonucleic acid (DNA) and the reproductive mechanisms of (DNA). Also, RNA-enzyme action was involved in acquiring DNA from neighboring bacteria, in DNA repair, and also in the rearrangement of DNA within the genes (Singleton, 1999).

Homo Sapiens, through eons of time, have evolved from one-celled bacteria to our present multi-cellular form. However, like other life forms, our bodies still contain the original obligate anaerobe archaebacteria that lived in the deep oceans of planet earth between 3 and 4 billion years ago. Archaebacteria, the oldest life form as yet discovered on earth, still live within hot thermal vents on the ocean floor. They ingest inorganic chemicals for nourishment and energy, and excrete oxygen, lethal to anaerobes, as a waste product.

In the early atmosphere of planet earth, the excreted waste created a build up of oxygen, and some archaebacteria evolved extra cell walls for protection from the lethal oxygen. The ozone layer, the protective covering that shields Earth from the dangerous ultraviolet rays of the sun, was created by the excess oxygen, and this allowed bacteria to emerge from the water and mineral deposits, and proliferate on land (Barghoorn, 1992). However, it appeared that bacteria still sought endosymbiotic relationships within other life forms.

Several questions arise. Why did bacteria proceed from an independent free floating existence to a dependent state in a larger life form? Why is it necessary for plant and animal life to contain trillions of bacteria? Is it possible that during the changing earth's atmosphere from a presence of less than 1% oxygen to approximately 21% oxygen, the anaerobic archaebacteria created increasingly complicated shelters to protect themselves in order to survive in a constantly changing atmosphere?

Further, as a result of the changing atmosphere, are all life forms on the planet Earth, including marine life, fungi, plants, animals, and humans, in reality those increasingly complicated evolved shelters? Have bacteria reached their ultimate goal? Have these microscopic creatures finally created a shelter in a human form that is now capable of producing the means of returning the atmosphere to less than 1% oxygen through pollution; an ever enlarging hole in the ozone layer; or a nuclear winter in which only anaerobic bacteria can survive to live in an independent state again?

Many people have a great interest in tracing their family lineage to understand where they came from, and/or they use psychological methods to understand how their families interact and behave. In a non-scientific survey carried out by this author, when asked what they knew about bacteria, the majority of answers given typically referred to the pathogenic species and the illnesses for which all bacteria are perceived responsible, ending in comments on how to eradicate bacteria in one's environment. Newspaper/magazine advertisements and television commercials are replete with products to eliminate bacteria. One could be led to the conclusion that only the scientific world is interested in tracing our original ancestors, bacteria, and discovering how and why the microscopic organisms interact and behave the way they do.

The gastrointestinal tract of a human is home to millions of bacteria involved in the digestion and absorption of nutrients, and also with the disposal of their wastes. However, in recent research it has been discovered that the digestive tract works independently from the central nervous system. In his book, *The Second Brain* (1998), Michael D. Gershon, M. D. describes the re-discovery of the independent enteric (gastrointestinal) nervous system. The author also describes his own personal research on serotonin, and the embryonic origin of the enteric nervous system in the gastrointestinal tract.

Davis and Solomon (1986) describe how the cranial brain developed later than the digestive system in invertebrate animals. Dr. Gershon (1998) names the enteric nervous system as the second brain. However, food digestion and waste expulsion were being carried out before a cranial brain was developed. Davis and Solomon (1986) proposed that the cranial brain first developed in the flatworm as an observation outpost in order to detect nutrients and avoid danger. Therefore it seems more likely that the enteric brain was the first brain, and the cranial brain gradually evolved and grew more complicated for the requirements of survival for each developing organism.

Types of Bacteria

Cocci.   Circular bacteria. Live in the body and are mostly harmless.

Bacilli.   Rod shaped bacteria. Usually exist independently but can join end to end.

Spirilla.   Rigid spiral shaped bacteria.

Spirocheta. Flexible spiral shaped bacteria.

Coccobacilli. Oval shaped bacteria.

Hyphae.   Threadlike bacteria which are called myceleum when they congregate.

Square bacteria and box-like bacteria are also to be found in the prokaryote world.

Definition of terms

actinomycetes - branched filaments in Fungi.

allogenic - forces from the environment acting on internal mechanisms of organism.

anaerobe - lives without oxygen.

anaerobic fermentation - energy is derived from sources other than
oxygen. e.g. E-coli use nitrates and fumerate.

angiosperms - flower and fruit bearing.

apoptosis - programmed death of a cell.

arbuscule - vesicle made by plant cell to enclose nutrients from mycorrhiza.

archaebacteria - ancient life forms - oldest living prokaryote
organism to which oxygen is lethal.

ATP - adenosine triphosphate.

autotrophes - obtain nutrients from inorganic substances. Make food from
CO_2 Require no organic food - use chemicals for growth.

chemoautotrophes - use chemical energy to make food.

chitin - nitrogen-rich polysaccharide of glucosamine units. Used
as an outer covering for support or shelter.

cuticle - lower level of shell covering.

cytoplasm - fluid within cells that contain the various vacuoles or organelles
required for nutrients, metabolism, and waste products.

ecdysis - molting or discarding the outer shells.

ecosystem - specific habitat of microbial community.

endosymbiosis - where two organisms of different species live together
with each entity providing benefit to the other.

enteric nervous system (ENS) - nervous system in the digestive tract.

enzyme - protein molecule that causes reaction in living cells - does not change form in the process - helps nutrient and energy metabolism.

epicuticle - upper layer of shell covering.

eukaryote - bacteria with defined membrane bound nucleus.

facultative aerobes - bacteria that usually grow without oxygen, but can grow with it.

facultative anaerobes - bacteria that usually grow in presence of oxygen, but can grow without it.

fermentation - acquiring energy without oxygen,

fimbrae - tiny appendages that bacteria use for attachment.

gymnosperms - evergreens.

glial cells - cells that lend support to neurons (nerve cells) by increasing their potential.

halophiles - bacterial that live in salt environments.

heterotrophes - require organic food (includes humans).

horizontal transmission - passed from a host of a different species.

invertebrate - organism with no spinal chord.

lipids - fats.

lithotrophes - obtain energy from inorganic compounds.

lumen - the area within the gastrointestinal tract where the digesta proceeds in a downward movement.

macrophages - white blood cells.

methanogens - convert CO2 to methane.

mycoplasmas - small bacteria with less DNA chromosomes than most bacteria.

mycorrhiza - (*myco* fungus + *rhizo* from the Greek meaning root) Symbiotic relationship between fungus and bacteria on plant roots.

neoteny - retention of the new.

neurons - brain nerves.

neurotransmitters - chemicals that transmit messages between brain neurons (nerves).

obligate aerobes - bacteria that require oxygen for maintenance and growth.

obligate anaerobe -   archaebacteria that do not use oxygen.

organotrophes - obtain energy from organic compounds.

photosynthesis - obtaining energy using sunlight.

preadaptation - existing structures changed to fit environment e.g. fins> limbs.

prokaryote - cell without a nucleus.

photo-autotroph - needs light to make food.

protozoa (sing. protozoon) - single celled heterotrophs e.g. amoeba.

pseudopods - foot like appendages.

respiration - acquiring energy using oxygen.

symbiosis -where two different species co-exist in one organism.

thermophiles - achaebacteria that live at hot springs.

vertebrate - organism with a spinal chord.

vertical transmission - characteristics handed down from female to female.

Chapter 2

In the Beginning.

Ponnamperuma (1992) postulates that chemical evolution preceded biological evolution. In its earliest beginnings, over 4.5 billion years ago, the planet Earth was a suspended, spinning, volcanic mass of molten chemicals. Among the boiling, bubbling chemicals were carbon, nitrogen, hydrogen, and oxygen; the four major elements present in every life form. It has been shown in many laboratory experiments that different combinations of these four chemicals replicate the formation of earth's earliest atmosphere of methane, ammonia and water.

In the studies of Miller and Urey, Honda, Navarro-Gonzales, and Ponnamperuma (as cited in Ponnamperuma, 1992) it was demonstrated that combining the earth's early atmosphere with the energy of radiation from the sun, and also with the energy from bombardment by asteroids, meteorites, and comets from space, the building blocks of organic matter, adenine, thiamine, guanine, and cytosine can be synthesized. These latter four bases, along with the backbone of phosphorus and sugars, form the spiral helix of deoxyribonucleic acid, more commonly known as DNA (Ponnamperuma, 1992).

It is proposed ribonucleic acid (RNA) preceded the formation of DNA. DNA requires the presence of RNA to reproduce, but RNA can replicate itself and produce its own proteins. Self-replication by RNA may be accomplished because RNA contains ribozymes acting with enzymes to produce proteins (Margulis & Sagan, 1995). Pollock (1994) says that because all living matter, through DNA, shares the same ancestry and chemistry, "we are related enough to a duck and an orange that we can eat them both" (p.13).

DNA is necessary for reproduction and contains all the codes for how the organism will grow and reproduce itself successfully. There are also instructions for self repair within the DNA sequence. In the days before the formation of the ozone layer, when ultraviolet radiation from the sun was a threat to bacteria, a protein named photolase was used with the help of DNA to repair any impairment (Singleton, 1999).

In order to reproduce, the DNA helix must be enclosed in an aqueous solution that is surrounded by a lipid membrane (Singleton, 1999). Hargreaves and Deamer (1978) have demonstrated that lipids and proteins can combine to form membranes which are then able to enclose the contents of a cell. The two scientists posit that these events could have occurred before life forms appeared on earth.

Although experiments have shown how a cell may have formed, there has been no link found that explains how bacteria came into existence. Three items are necessary for life as we know it; energy, water, and organic molecules that contain carbon. These three items are found in abundance throughout the universe. Mercury, Mars, and the Moon have ice at their poles. Many outer planets have moons that are covered in ice. Comets are vast icebergs that contain water and organic compounds. The molecular clouds that surround our galaxy contain ice particles that could help form organic molecules (Feferman & Stobie, 1999).

Although there was an abundance of water in the early days of the earth, oxygen was not free. The oxygen available to us now, was released from the water, and added to the energy from the sun by the photosynthetic abilities of the prolific cyanobacteria and plants. Energy can be supplied by the central star of a galaxy, e.g. our sun, but energy on earth can also be supplied from the its hot, molten center in the form of hydrogen sulfide, used by thermophylic bacteria and a myriad of sea creatures that live, eat, and multiply on the ocean floor, without aid from the sun, (Feferman & Stobie, 1999).

Neil de Grasse Tyson, the narrator in the PBS television show *Origins: How life began* (Harper, 2005), commented on the meteorite which landed in Murcheson, West Australia. The meteorite contained amino acids, the building blocks of life. Seventy different amino acids have been found in meteorite particles. The television show also presented an experiment employing high velocity impact that demonstrated how meteorites, containing the four inorganic ingredients needed for life, carbon, hydrogen, oxygen, and nitrogen, came crashing to earth at high speeds. The energy expended on impact on earth's early surface, transformed the necessary four ingredients for life into amino acids, which also occurred in the experiment. Amino acids are the precursors of peptides, which join together to form protein that is present in all living cells.

Recently, organic molecules that are between 4.5 and 5 billions years old have been collected from the edge of earth's atmosphere. Many theories embrace the idea that organic molecules came from outer space, and their counterparts arrived on earth transported by meteorites and comets (Harper, 2005). Some theories include the possibility that bacteria came to earth from space; enclosed within the safety of meteorites' 40 degree centigrade centers (Kimball, 2004). However, no one has been able to find the link that transforms inorganic matter to organic life.

Bacteria

In an amusing and engaging story, *The Other End of the Microscope,* Koneman (2002) describes the life and times of bacteria - from their point of view. In the book, delightfully illustrated by Bert Dodson, the bacteria are given the opportunity to say how they perceive the encroachment of humans on what was originally their territory. They recount how, through millennium after millennium, the bacteria remained undisturbed and undiscovered until a 16th.century Dutch biologist by the name of Antonie van Leeuwenhoek peeked into his magnifying lens and saw hundreds of tiny creatures swimming back and forth. Amazed at the sight, van Leeuwenhoek called the little creatures animalcules. Louis Pasteur went on to investigate the prowess of bacteria, and Robert Koch later concluded how virulent bacteria could evolve and cause human diseases. In their discourse in the book's narrative, the miniature marvels eschew how humans invented technologies and acquired destructive ways of destroying the environment, to the detriment of bacteria. Yet the bacteria espouse, "through centuries of gene transfers, chromosomal recombinations, and mutations" (Koneman, 2002, p. 7), they have survived and multiplied, and so have been able to overcome the onslaught of human proliferation. An admonition from the bacteria to humans is that they need to be more aware of the role bacteria play in the survival of humanity and the planet.

A bacterium is a one-celled prokaryote with a membrane cell wall containing a solution called the cytoplasm, within which are enzymes, ribosomes, and storage units for nutrients or waste products. The cytoplasm also contains the mechanisms for supporting cellular functions, self maintenance, and self replication. Many bacteria have outer walls of proteins and lipids, some have no walls, e.g. Thermoplasma, an extremely ancient archaebacterium that lives in hot, acidic environments (Margulis & Schwartz, 1998).

Bacteria are considered one species because they can interchange their DNA. Only nucleated organisms, which are products of symbiogenesis, merged genomes, can be classified as different species (Margulis & Sagan, 2002). Further, the two scientists refer to the bacteria as different *strains.* For optimum growth and reproduction, the bacteria must have a good supply of nutrients, energy, necessary temperature, pH level, and oxygen requirement for their particular strain. Bacteria also need to have water inside and surrounding the cells (Singleton, 1999).

A distinguishing feature of a bacterium that differentiates it from the cells of all other organisms is that a prokaryote cell has no nucleus. The DNA of most bacteria is usually contained in a single, multi-folded chromosome thread (nucleoid) in the form of a closed loop, but in some bacteria the chromosome thread may be linear.

The codes dictated by DNA chromosomes pass on the behavior and traits to the next generation. To replicate, most bacteria engage in asexual reproduction through binary fission, where the DNA attaches to the cell membrane, the cell divides in half with the result that both replicated cells are identical. Bacteria have no gender gap. Having no sex cells, there are no male or female bacteria (Koneman, 2002). There are about 20,000 strains of bacteria (Margulis, 1992). Transformation enables genetic exchange when broken bits of DNA float around in the aqueous environment and are taken up by passing bacteria. The pieces of DNA are imported into the genes and are passed on to progeny. For movement, some bacteria just float around in the surrounding medium. Other bacteria have flagella, hair like appendages, that beat swiftly for propulsion, and others secrete mucus upon which extended pseudopods enable mobility (Davis & Solomon, 1986).

Research shows that the amazing longevity, resilience, and survival rates of bacteria is due in part to their ability to adapt to their environment. If a new presence affects their survival, the change is incorporated into their DNA. A few progeny survive that can deal with the change, and then the change is passed down to their progeny (Singleton, 1992).

There are billions of bacteria to be found in soil, and as a measure, Postgate states that there are "200 to 500 pounds of microbes to every acre of good agricultural soil" (p. 3). Further, the population of bacteria makes up about 90% of all the living material on our planet. Bacteria play an extremely important role in the health and viability of the planet by breaking down, and then recycling, the organic molecules of dead vegetation and the inorganic molecules of metals. Without the recycling abilities of bacteria, the planet would be overwhelmed and life for evolved organisms would not be possible (Singleton, 1999).

Bacteria are the most successful organisms having survived and multiplied for billions of years. When nutrients and moisture are scarce, the bacteria produce enzymes that seek out nutrients from other sources, or produce antibiotics to destroy competitors. If the community becomes too large for available nutrients, some bacteria produce a toxin that causes their own demise (apoptosis). In extreme shortages, bacteria create thick protective shells around themselves, and they stay dormant until their surroundings improve to a state in which they can survive and multiply again. After thousands of years, the bacteria can re-activate themselves when food and water become available (Tannock, 1995).

The oldest known form of life are bacteria. Life forms are divided into five kingdoms named Monera, Protoctista, Fungi, Plantae, and Animalae; humans are included in the last category (Margulis & Schwartz, 1988). Bacteria, are members of the Monera family. Carl Woese (as cited in Margulis & Sagan, 1995) showed that similarities in RNA structure divided the bacterial world into two divisions. The most ancient bacteria which live without oxygen in extremely hot, acidic or salty climes are now called archaebacteria. The other bacteria named eubacteria (Snedden, 2000) are mostly referred to as just bacteria.

Plasmids

Most bacteria have additional small strands of circular or linear DNA floating around in the cytoplasm. The floating strands are self-replicating plasmids and conjugation is applied, whereby, through a connecting pilla or tube, DNA is passed from one plasmid to another leading to variety which promotes survival in changing environments. Plasmid conjugation enables genetic exchange between the same bacteria, related bacteria, or even unrelated bacteria. It is also possible for some plasmids to dictate whether the invaded cell will be male or female, or change an invaded female cell to male The tiny plasmids pass easily in and out of bacterial cells (Postgate, 1992).

Some plasmids are able to carry out several duties, and there may be more than one kind of plasmid in a bacterial cell. Plasmids, named episomes, merge and then replicate with the host DNA, producing a system that over time negates antibiotic treatment. Col- plasmids can manufacture proteins that are used to kill other bacteria. Degrative plasmids help to break down and digest difficult materials, and virulence plasmids convert bacteria into pathogens (Wikipedia, 2005).

Carbon is designated as the ingredient necessary to define life. Archaebacteria, called autotrophs, consume inorganic materials to grow and replicate. Most other bacteria, heterotrophs, require complex carbons for nutrition and cellular metabolism. Many cells can operate on one carbon source, others require several carbon sources (Postgate, 1992).

Magnetotactic bacteria have magnetic parts that line up with the north and south poles that enable the bacteria to move to deeper waters away from oxygen (Clement, 1996), and towards more conducive surroundings (Snedden, 2000). *Deinococcus radiodurens* bacteria survive at extremely high levels of radiation. They use the same mechanisms that are involved in dehydration that enable them to survive in times of energy and nutrient shortages. These hardy bacteria can also repair their own DNA (Leadbetter, 1996).

Pathogenicity

The majority of bacteria in the human body are healthy and carry out important and necessary functions. Some strains of bacteria, however, can become pathogenic if given the right circumstances, and these micro-organisms are called opportunistic pathogens, e.g. if there is a lesion in the intestinal lining, normal bacteria passing through to the bloodstream can become pathogenic (Singleton, 1999).

Viruses

Viruses are not generally considered by scientists to be a life form. They require a living cell to reproduce themselves. A virus is a small particle enclosed in a protein capsule (capsid). Each virus contains only a few enzymes and one strand of DNA or RNA, never both. Extra pieces of bacterial DNA are passed from one bacterium to another via viral plasmid conjugation or by protein coated viruses. Some viruses assemble a further coat (envelope) from the membranes of cells under attack, and then the new protein coat is used for attachment and entry to the targeted cells (Snedden, 2000). The attached viruses drill holes through the membranes, and then pass the DNA or RNA molecules into the cytoplasm of the cells. The viral strands then use the reproductive systems of the cells to replicate themselves. Some viruses kill the cells by breaking them open for release into the surrounding environment, while other viruses let the bacterial cells live, but cause them to keep producing new viruses which they must release into the surrounding milieu (Singleton, 1999).

Bacteria exposed to ultraviolet radiation shatter into numerous viruses which spread disease and infection (Margulis & Sagan, 1995). The rabies virus gains access to the brain where it is safe from attack from the immune system. The virus causes aggression in the animal, which then attacks and bites another animal, and the virus is passed along in the saliva to another host (Sapolsky, 2003). A virus that invades a bacterial cell is called a bacteriophage.

Prions, depending on their shape. can either be pathogenic or not. They are infinitely smaller than viruses and consist solely of protein. Usually not considered a life form, prions contain no nucleic acid. Slow acting while attached to nerve cells, pathogenic prions eventually kill the brain cells. Loss of muscle control and eventual death is the result. Scrapie in sheep and Mad Cow Disease in cattle are tragic results of prion invasion (Snedden, 2000).

Rickettsia are the smallest bacteria. They can cause fever and rash. Lice are the vectors for the miniature menace, *Rickettsia prowazekki* (Singleton, 1999).

Pathogenicity Islands

Sequencing of the bacterial DNA genome has led to intensive investigation of microbial evolution. Evolution, until recently, has been studied on the long term inheritance expression of genes for the purpose of improving the organism. However, with the advent of modern technology and the sequencing of the human genome, it is now possible to study the action of small segments of genes, and their effect on the evolution, or effect on the pathogenicity, of life forms. In genetic variation, there are evolutionary factors that can encourage diversity, and other genetic factors that hinder the process, and then there are factors that can either help or hinder. Nucleotide sequences may be altered, sections of DNA may be moved around, or sections of DNA may be incorporated that have come from another similar strain of bacteria (Arber, 2002; Arthur, 2002).

Pathogenic bacteria contain pathogenicity islands, healthy bacteria do not (Hueck, 1998). Pathogenicity islands are sections of DNA that can change a bacterium from a nonpathogenic state into a pathogenic entity. The re-arranged genes of the pathogenicity islands can already be part of the host genome, may have been deleted from the host genome, or they may have been incorporated into the host genome by horizontal acquisition, i.e. the pathogenic section of DNA is transferred from a similar strain of bacteria via a plasmid or a bacteriophage (Kingsley & Baumler, 2002).

Genomes evolve not only by acquisition of new pieces of DNA following gene transfer, but can transpire by loss of genes. Functional loss in the genome of *Yersinia pseudotuberculosis* led to the emergence of an extremely virulent *Yersinia pestis* (Carneil, 2002). *Shigella,,* the cause of diarrhea, appears to be a shortened genome version of an *Escheriscia coli* strain, the latter able to stop the infection (Ingersoll, Groisman, & Zychlinsky, 2002). Only *E. coli* equipped with fimbriae can attach themselves to host cells, and they sometimes obtain fimbriae from bacteria other than their own strain (Koneman, 2002).

A smaller version of a non-pathogen *Bordetella bronchiseptica,* named *Bordetella pertussis*, causes whooping cough. It is posited that the loss of genes in the *Bordetella* genus may have been instrumental in the evolution of their pathogenicity (von Wintzingerode, Gerlach, Schneider, & Gross, 2002). The passing of pathogenicity island genes from one bacteria to another is thought to have been instrumental in the evolution of pathogenic bacteria (Groisman & Ochman, 1996). It is thought that pathogenic bacteria evolved further to encompass evolving organisms, e.g. *Salmonella enterica* sub species 1 probably evolved from infection in reptiles to include infection in warm blooded animals (Ochman & Wilson, 1987).

Small, excess pieces of DNA that cross from one chromosome to another are called transposons. Transposons may also enter plasmids, and when the plasmids cross into other bacteria, they become part of the genetic makeup of the invaded bacteria. In some instances, this genetic incorporation can either make the new bacterium non-viral or make it pathogenic (Koneman, 2002).

Many strains of bacteria are called opportunistic because they can take advantage of certain situations. *Pseudomonas aeruginosa* invades in cases of lowered immune systems, and *Burkholderia pseudomallei* in wet, moist climates causes a foot rot that can heal, and then reoccur many times over (Koneman, 2002).

Transport Systems.

Transport systems that carry vital factors in and out of the cytoplasm, e.g. energy in - wastes out (Singleton, 1999), can be used by pathogenic bacterial cells to deliver virulence factors. There are four secretion pathways, numbered I through IV, that the bacteria use to carry out their missions. The Type 1 pathway is used to transport ATP enzymes for energy, and also proteins that are capable of fusing and building cell membranes (Hueck, 1998). Pathway Type II conveys enzymes that will be used to break down material in the cytoplasm. The Type III secretion system (TTSS) is a virulence transport pathway that pathogenic bacteria use to convey pathogenic proteins into the cytoplasm of eukaryote cells. Many pathogenic bacteria contain about 20 genes that are responsible for creating the virulent Type III secretion mechanism. Enzymes also use the Type IV pathway, but their mission is to cut a hole in the membrane so that the pathogenic proteins can pass into the host cell. Some pathogens adhere to the outside of the targeted cell, and use the transport systems to conduct proteins and enzymes involved in the attack. By employing its Type III secretion system, the bacteria operate their virulence factors from the outside surface of the besieged cells (Torres & Kaper, 2002). In the Type III secretion system, chaperones attach themselves to the proteins to ensure there is no interference from other proteins before the site of action is reached. There is a fifth transport pathway that is used to convey plasmids (Singleton, 1999).

In most instances of invasion, however, the bacteria instruct the host cells to assist in their uptake. When pathogens attach to the surface of a host cell, the host cell assists the pathogens to enter the cell by conforming its membrane structure to allow entry, and then enclosing the invaders in vacuoles. After the invading bacteria are contained in a vacuole, inflammation occurs when blood is sent to the area accompanied by macrophages, the body's defenders. In the ensuing battle between the pathogens and the macrophages, a great deal depends on the health and strength of the host. In a healthy body, membranes release a mucous, containing its own antibiotics, that bathe the tissues, and so prevents adhesion, a strong immune system provides defensive mechanisms, and the established healthy bacteria prevent the invaders from gaining a foothold (Singleton, 1999).

In a lowered immune system, the pathogens replicate inside the vacuoles, and those with cutting enzymes can eventually escape into the circulating pathways in the body. There are also strains of bacteria that can invade, and kill, the macrophages that are part of the defensive mechanism of the body's immune system. Some macrophages even allow bacterial invasion by failing to initiate their defensive mechanisms (Singleton, 1999).

Pathogenic bacteria, as shown by Koneman (2002), use toxins and enzymes in their invasive repertoire. Toxins can be released that are able to destroy the body's blood and tissue cells. *Streptococcus pyogenes* use proteolytic enzymes to allow free access to tissues that are then destroyed. In cases of illness involving the "lymphoid cell system" (p. 95), invading bacteria attach themselves to the antibodies that were released to destroy the invaders. Later the pathogens return and maybe cause a more serious and debilitating illness. Again, similarities between the membrane structures of the body cells and the pathogens allow easier penetration for the destructive bacteria.

Several pathogens can cause disease instantly in an organism. Others can lie suppressed by the immune system, and then can be passed on to other organisms without the carrier suffering any ill effects. In some cases the disease can be caused by infected food or water. The neurotoxin pathogen, *Clostridium botulinum,* found in contaminated foods, can cause extensive nerve damage and death (Singleton, 1999).

For a defense mechanism against the immune system, pathogenic bacteria, e. g. *Streptococcus pneumoniae,* hide in thick, polysaccharide capsules that protect them from attack by white cells. After cell invasion and replication the pathogens escape. Flooding the body, the pathogens attack the essential organs causing tremendous inflammation. It takes a considerable amount of time for enough special antibodies to accumulate and be strong enough to remove the capsules protecting the rampaging pathogens (Koneman, 2002). Some polysaccharide coated pathogens, e.g. *Haemophilus influenza* type b, *Neisseria miningitidis,* and *Streptococcus pneumonias* can be destroyed by a competent immune system in adults, but cause infections in young children because of insufficient antibodies (Singleton, 1999).

Mycoplasmas, up to now, are the smallest bacteria discovered. Without walls, they are immune to antibiotics which only work on cell walls. Responsible for many infections in plants and animals, they cause few diseases in humans (Snedden, 2000).

Unless there is an open wound, the biggest obstacle for pathogenic bacteria, when trying to colonize an area, is the presence of resident microflora. Established communities of bacteria are equipped to repel unwelcome invaders. However, if bacterial communities are depleted, successful, pathogenic invasion is possible. Antibiotic treatment increases the chances of a depleted microflora. In the gastrointestinal tract, diminished bacterial colonies can be catastrophic. Complete digestion, absorption of nutrients, the ability to synthesize essential vitamins, and the regular evacuation of wastes is essential for a healthy body. All of these necessary tasks are carried out by the numerous communities of resident bacteria. *Clostridium difficile* is an especially potent bacterium. It secretes two toxins that act together to penetrate the lining of the colon and so create a passageway for *C. difficile* to cross into the once protected sub mucosa. All animals, including humans, when exposed to stressful situations have a greater chance of depleted bacterial communities in their digestive tracts than animals not under stress. A compromised immune system may also be a result of stress, and this can increase the chances of infection (Tannock,1995).

The immune system is a barrier of defense against pathogenic intruders. One defense strategy is the release of proteins or enzymes that inactivate the pathogenic bacteria. A noticeable defense is the creation of inflammation around the infected part. In inflammation, blood suffuses the damaged area providing a protective barrier to prevent further penetration, and also the blood allows fast and easy access for immune factors to reach the site of infection, destroy the pathogens, and allow for healing (Singleton, 1999).

Another defense mechanism is carried out by phagocytes. The pathogen is enclosed in a membrane, the pH is lowered, and enzymes degrade the invaders (phagositosis). However, some of the bacteria survive, replicate, then lyse the enclosing membrane and escape. Other pathogens provide themselves with protein or polysaccharide coatings that allow them to hide from the immune system defenders. Special proteins provided by the invading bacteria can also attack, invade, and kill the cells of the immune system. Antibodies are formed by the immune system to attach and then attack specific bacteria, but by altering their membrane surface, the disguised micro-organisms escape detection. Certain pathogens can interfere with the host cells communication systems. Some can provide toxins that can bind and kill, or some can alter their gene sequences in order to provide different surface proteins (Singleton, 1999).

With the aid of enzymes, the virulent *Staphylococcus aureus* can carve out a pocket in an organism's tissue, and then line the pocket with an impenetrable wall. While the pathogenic bacteria are filling up the carved out space, they release proteins to subdue or repel the body's defense systems. In a sick room or hospital setting, *Staphylococcus epidermidis*, an especially virulent bacteria, can exude a sticky slime substance that allows a build up of the pathogens on medical instruments. Pathways are made through the slime for nutrients to pass to lower level bacteria, and then the pathways are re-formed to excrete wastes. The slime also acts as protection from any attempt to neutralize or destroy the pathogenic build up. The safest way to get rid of the virulent bacteria is to dispose of the invaded instrument (Koneman, 2002).

Human disturbances of bacterial ecosystems have led to infections and diseases. An outstanding example is the excavation of the bacteria *Legionella pneumoniae* from their watery depths of a nearby lake. The resurrected bacteria were then pumped, along with the water necessary for air conditioning, to the hotel where members of the American Legion were staying. In the air ducts of the hotel, the visiting bacteria found warm damp niches where they could rest and multiply before exiting through the air vents into the rooms of the sleeping Legionaires. Many of the Legionaires became ill, and unfortunately some of those who fell ill, died. It was many weeks before the cause of their deaths was discovered, and it was the bacteria who were blamed for the mishap. Should the real culprit have been the act of disturbing a natural native population which had occupied the waters of the lake undisturbed for eons of time (Koneman, 2002)?

Animals under restraint become stressed, and when this occurs, residential, protective bacterial communities are reduced allowing pathogenic bacteria the opportunity to attach and proliferate in the digestive tract (Tannock, 1995). Also, there are temperate phages (bacterial viruses) that lie dormant, and only become actively virulent when stress occurs (Postgate, 1992). Cattle, who used to roam in open fields, are now penned and restrained to enlarge beef production, and, while penned, they are fed harvested grain. The result, grain-fed cattle ready for slaughter were found to contain the bacteria E. coli 0157.H7 which are acid resistant, enabling the highly pathogenic bacteria to travel through the stomachs of the cattle without being degraded by the stomach acid. The virulent bacteria when passed on to the human population cause disease and death. Feeding hay to the cattle for five days before slaughter greatly decreased the number of E.Coli 0157.H7. Cattle raised in open pastures were found to be free of E.coli 0157.H7 (Segelken, 1998).

Most bacteria can double in twenty minutes. Koneman (2002) points out that any attempt to totally eradicate pathogenic bacteria will end in failure. Bacteria can reproduce and mutate so quickly that just one lone survivor, with its own "unique self sustaining mutation" (p.98), can multiply and ensure adaptation to any adverse situation.

Bacteria in Evolving Organisms.

Bacteria

Necessary and mutually beneficial relationships occur between bacteria and all life forms. Nitrogen fixation is an extremely important and necessary relationship for building proteins. Certain bacteria have the ability to take nitrogen, the most voluminous gas surrounding our planet, to break its triple bonds, and then to combine it with organic substances to make the needed proteins (Margulis & Sagan, 1995).

In primordial times, bacteria fed on simple sugars that were obtained from the elements around them. As bacterial count increased, however, nutrient availability became scarce, and some strains of bacteria, e.g. cyanobacteria, merged and developed ways of making food from the sun's energy (photosynthesis); at the same time acquiring the ability to store energy for future use. Cyanobacteria contain plastids, which provide the ability to photosynthesize. The progenitor of plants are cyanobacteria (Margulis & Sagan, 1995).

First named blue/green algae, cyanobacteria developed a method to split water molecules, absorb the hydrogen, and excrete the oxygen as a waste product, which, over millions of years, helped change the earth's atmosphere from less than 1% oxygen to 21% oxygen. The adaptation to photosynthesize from the energy of the sun allowed cyanobacteria to proliferate over land and ocean surfaces (Harper, 2005).

The processes which enable us to incorporate oxygen for our energy needs has been passed down to us from bacteria. Oxygen protects the earth from harmful radiation from the sun by creating an ozone layer in the upper atmosphere. The ozone protective cover allows for the creation and support of multiple organisms that live in the sea, in the air, and on land. It is posited by many scientists that breaks in the ozone layer could cause drastic changes in the Earth's land and ocean configurations. These changes could lead to disastrous results for many life forms (Harper 2005).

In the past, mats of cyanobacteria formed reefs and barriers around coastal waters. Presently, mats of living bacteria float along shore lines and are made up of roughly three layers. The lower layers of bacteria use low density light, the middle layers require sulfide from the organisms beneath and sunlight from the photosynthesizers occupying the layers above (Margulis & Sagan, 1995).

Fortey (1997) shows how large, layered mats of slime-covered ancient bacteria can still be found in hot, moist tropical areas, and he posits that the bacteria from over 3,500 million years ago may still be there. It is therefore possible that the bacteria are in a quiescent state, and they could revive when conditions become conducive for their survival, e.g. an atmosphere of less than 1% oxygen.

On land, cyanobacteria combine with fungi and attach to rocks as lichen, making it the first resident of each land-based ecosystem. The lichen breaks down the rocks into soil, and the cyanobacteria provide the nitrogen used in making proteins for growing plants. Plants, in turn, provide the necessary food and energy for other life forms..

Protoctista

The second basic cell of life is the eukaryote cell, in which the DNA is enclosed in a nucleus. Protoctists are eukaryote cells. Margulis and Sagan (2002) suggest that "Metabolic dialogue and physical proximity led to incorporation, accommodation, and rearrangement" (p. 97). It is now generally accepted that eukaryote cells evolved when bacterial communities became too large for available nutrients. Different strains of bacteria joined forces and were able to exploit their different capabilities in order to survive.

Eukaryote cells are the result of bacteria ingesting, but not digesting, other bacteria, and so merging. Although one organism appears to be eating the other, the ingested organism continues to live, and genetic fusion of both organisms can be the result. In fusion, over time, it is ingestion, not digestion, of other organisms that leads to merged genomes, fuels evolution, and creates diversity of species. Further, the merged bacteria formed cells that had their DNA enclosed in membrane structured nuclei (Margulis & Sagan, 2002). In evolution, symbiosis occurred in "serial symbiosis...refers to acquisition of particular symbionts in a certain order" (Margulis,1992, p.151). In his book, *Stages of Evolution,* Brace (1998) points out:

> The nucleate cell is the basic building block for all plants and animals. The mechanics
>
> of cell division, the phenomenon of sexual reproduction, the control of heredity by
>
> DNA, and the roles of structural and enzymatic proteins are all essentially the same.
>
> (p.44)

Therefore, on the basis of the merged bacteria, the formation of the enclosed nucleus, and the same natural tendencies of reproduction, the evolution of all life forms up to the present day have been accomplished (Margulis & Sagan, 2002).

Eukaryote unicellular and multicellular protoctists, varying in size from amoebas to kelp, are members of the Kingdom of Protoctista, and they all live in a liquid based environment. The cells enclose different kinds of organelles, e.g. mitochondria to produce oxygen based energy, or chloroplasts for sun based energy. Some organisms float in water or wind currents but others have hair-like appendages that they can beat or rotate to achieve direction. Protoctistas can live as independent organisms or, by releasing certain chemicals, can join together to build larger structures. Waterborne protoctistas, in the form of plankton and algae, serve as food for marine life (Margulis, 1992).

Reproduction in protoctists is carried out in two ways, either by division of the cell in asexual reproduction, or by exchanging nuclei through sexual reproduction. In their book, *Mystery Dance: On the Evolution of Human Sexuality,* Margulis and Sagan (1991) posit that protists (one-celled microscopic protoctists), through meiosis, were the initiators of biological sex by "the pooling of two genetic systems each with its own DNA nucleotide strands to form a genetically distinct individual" (p.182). Also, they state that, meiosis, in which haploid cells have to be fertilized by egg and sperm, has been passed down through the ages from protists to its progeny, which includes humans. Further, this process meant that eukaryotes were the first cells to age and have programmed death.

During the life of a human, each cell usually divides between 80 and 90 times. Within most other organisms, cell division occurs about 50 times. For each cell division in eukaryotes, the separation of chromosomes is controlled by telomeres that become shortened with each division. Eventually the telomeres disappear leading to death of the cell. All descendents of eukaryotic protoctista, which include all evolved life forms, age and eventually die. Bacteria, on the other hand, never die through aging. They do not possess telomeres, and reproduce through binary fission; cell division which produces two identical cells (Snedden, 2003).

If living conditions become overcrowded, some bacteria will self sacrifice by releasing autotoxins in order for the rest of the community to survive.

Chloroplasts and mitochondria, organelles within eukaryote cells, are said to be the results of bacterial union; the ribosomal RNA of both cell organelles appear to be that of prokaryote cells. Both kinds of organelles contain their own DNA and replicate separately, but replication takes place at the same time as DNA replication in the cell nucleus.

Chloroplasts, evolved from cyanobacteria, are found in algae, protoctists, and plants, and use photosynthesis for energy. Chlorophyll in the cells captures sunlight, combines it with carbon dioxide and water, so providing nutrients for the organisms.

In *Car Talk*, an episode of the television series Science Now (2005), the power of hydrogen propelled cars was discussed. However, the show noted, to produce pure hydrogen is very expensive. An alternative proposed solution is to separate the hydrogen molecules from the oxygen molecules using the abundant commodity of water. The aim is to produce an energy product and, at the same time, to have a non-threatening environmental residue. This procedure is extremely difficult with today's technology, and so the price of separation is prohibitive (Cort, 2005). The irony of the problem is that bacteria learned to perform this procedure millions of years ago. The procedure is known as photosynthesis. In photosynthesis, energy from the sun, with the aid of chlorophyll, is used to separate the two hydrogen molecules from each oxygen molecule in the water. The hydrogen is used as energy for the bacteria or plants, and the oxygen is either used by the organisms for respiration or it is released as a by-product.

Mitochondria organelles are a combination of bacteria that came together to produce energy by making use of the ever increasing levels of oxygen in the environment. Like bacteria, mitochondria reproduce similarly by binary fission In contrast to fermentation in anaerobic energy production where only two molecules of adenosine triphosphate (ATP) are produced from one molecule of sugar, in oxygen respiration, thirty six molecules of ATP are produced (Margulis & Sagan, 1995). Mitochondria are a part of all respiring cells in evolved organisms.

Slime Molds

A myxomycete is a one celled eukaryote organism better known as slime mold (Stephenson & Stempen, 1994). According to Davis and Solomon (1986), there has been difficulty in classifying it because of its "Animal or protozoan characteristics during part of life cycle; fungal traits during remainder" (p. 260). Some scientists have classified it as protozoan and others as fungi (Snedden, 2000). A myxomycete, at one stage of its life, is a mobile protoplasm called a plasmodium (plural, plasmodia), that can creep along while ingesting organic substances and bacteria (Stephenson & Stempen, 1994). Slime mold can identify the number of bacteria present by molecules the bacteria release. Cyclic AMP (adenosine 3', 5'-monophosphate) signals, that show bacteria are present, are then sent to other myxomycetes in the area (Snedden, 2000). Plasmodia are also known to ingest other plasmodia, leading to cell fusion. As it grows, the myxomycete moves to a drier area where it produces spores. The spores, which contain bacteria, release cells which can merge with other compatible cells to produce zygotes that grow into plasmodia. If the cells are released in areas not conducive to growth, the cells form hard shells and become dormant until conditions improve (Stephenson & Stempen, 1994).

A plasmodium also has the ability to produce a hard shell coating in unsuitable environments, later to revive when food and water become available. Some myxomycetes are found in aquatic environments but most have been found in the company of bacteria among decaying and rotting trees or vegetation. They are also found in the company of many bacteria in the fertile top soil of forests, fields, or cultivated areas, where the fungi and bacteria surround the roots of vegetation promoting healthy growth (Stephenson & Stempen, 1994).

Fungus

Sometimes mistaken for plants, fungi (sing. fungus), occupy the Kingdom of Fungi. The fungi, which include molds (zygofungi), yeasts (ascofungi), and mushrooms (basidiofungi) which according to Margulis (1992), are descended from "ancient transparent protoctists" (p. 37). Along with bacteria, fungi are the earth's recyclers of dead and rotting vegetation. Fungi develop from single spores that contain bacterial spores For structural support, fungus have chitin walls (Stephenson & Stempen, 1994).

Lacking chloroplasts, fungi cannot photosynthesize. When large enough, fungi secrete enzymes outside of their bodies to digest nutrients from the plants or animals to which the fungi have attached themselves. The enzymes break down the food molecules which are then able to pass through the outer membranes of the fungi cells. To discourage would-be invaders of their territory and food supplies, fungi emit powerful toxins (Hudler, 1998).

Hudler (1998) shows how fungi, along with bacteria, are the ultimate recyclers of dead and decaying materials. When, for example, a live tree branch is broken, bacteria first enter the break and build a community. Then, joined by fungi, they both begin to break down the internal structure of the tree. Further, plants which have no chloroplasts in their cells, and therefore cannot photosynthesize (e.g. the plant Monotropa), have an endo-symbiotic relationship with fungi and bacteria, which provide water and nourishment through the roots of the plants. Fungi can also protect plants from invading pathogens.

Most vegetation, including trees, shrubs, and plants, share their root systems. Mycorrhizae on the plants root systems provide nutrients and moisture. To a lesser degree, the fungi and bacteria receive nourishment from the plant, but they are assured protection in hazardous soil conditions (Hudler, 1998). Each plant sharing its root system with like family members enable plants to colonize desolate areas, and when the earth was young, allowed for plants to spread and inhabit its barren lands (Davis & Solomon, 1986).

Lichens are an endo-symbiotic relationship between cyanobacteria and fungi, the fungus making up the larger portion of the association. Fungi can grow on bare rock in barren environments and so protect the cyanobacteria from harsh and hostile conditions. In return, the cyanobacteria provide sunlight, nourishment, and energy for the fungi. Extreme hazardous conditions such as scorching heat or drought conditions, however, do not inhibit the growth or dissipation of lichen via spores (Hudler, 1998). .

Dispersal by the wind of spores or flakes of fungi containing bacterial spores, ensure that the lichen will grow again in near or far locations (Hudler, 1998). Lichen become the precursors for an ecological niche of plants. Due to its ability to cling to bare ground or rock, and its ability to break down the rock to soil, lichen, a combination of many nitrogen fixing bacteria and assorted fungi, can form ecological niches and introduce vegetation in barren or deserted areas (Postgate, 1992).In areas of sparse habitation, lichen may serve as food for animals and humans (Hudler, 1998).

Plants

Davis and Solomon (1986) show how plants evolved from green algae (now named cyanobacteria), and acquired the ability to conserve water on the inside and build a strong outer shell of lignan and cellulose on the outside, enabling the plants to survive on land. Further, bryophytes, consisting of mosses and liverworsts, are more highly developed than algae. They have tissue structures, but are non-vascular so their root-like systems are grounded in moist areas to enable fertilization, which is accomplished by alternating generations of asexual (producing spores)and sexual (producing gametes) methods.

Cyanobacteria are credited with being the progenitor of tracheophytes. Tracheophytes developed a vascular system with stems and roots to support a system that could acquire and disperse water, nutrients, and minerals. The flowers of the plants enabled fertilization and reproduction. As stems became stronger and longer, shrubs, and then trees with bark for support and protection, developed a method for storing water. These plants are called gymnosperms and angiosperms (Davis & Solomon, 1986).

Engaging in sexual reproduction, all vegetation grows from a sperm/egg fertilized embryo. As in the animal world, after fertilization, the embryo grows within the female plant. The embryo develops into a sporophyte plant, which, in turn, produces spores that become gametophyte plants which can be either sperm-producing males or egg-producing females (Margulis, 1992). Plants with seeds ensured that the plant within the hard case-covering could survive until growth conditions were optimum (Davis & Solomon, 1986).

Nitrogen fixing.

Bacteria are found on all root systems of plants, shrubs, and trees. Some strains of bacteria take up phosphorus for the plants. Other bacteria are responsible for extracting nitrogen from the air and supplying it to the plants. Nitrogen is necessary for building proteins for growth (Margulis & Sagan, 1995). Plants cannot absorb nitrogen gas from surrounding air. Nitrogen-fixing bacteria, which occupy the root nodules of plants, employ the anaerobic enzyme nitrogenase to catalyze nitrogen fixation (Davis & Solomon, 1986). A great deal of ATP energy is required for the process and apparently only bacteria can provide nitrogen fixation quickly and efficiently (Singleton, 1999). Davis and Solomon (1986) go on to say that "these mutualistic bacteria can fix nitrogen 100 times faster than other, less vigorous nitrogen fixing organisms" (p. 406). In many countries of the world, food production relies on the abilities of nitrogen fixing bacteria (Postgate, 1986). Animal life on land relies on vegetation to provide its food. Meat-eating animals eat the digestive systems of plant-eating prey first to obtain vegetation nutrients.

Animals

Invertebrates make up about 95% of the animal kingdom. Probiscus worms (Phyla Nemertinea) possess the first rudimentary digestive system having "a tube-within-a-tube body plan...with a mouth at one end for taking in food and an anus at the other for eliminating wastes" (Davis & Solomon, 1986, p. 316). Vertebrates, which include fish, birds, and animals, eventually developed. Fossil evidence appears to support the idea that life on Earth originated in the oceans, and gradually some bacteria and evolved creatures moved to occupy land (Fortey, 1997).

To qualify for inclusion in the Kingdom of Animalia, a multi-cellular organism must develop from an embryo, the result of a sperm-fertilized egg (Margulis,1992). Animals are popularly thought of as land dwellers, but the vast majority live in fresh or sea water. Occasionally, fish or sea creatures thought to be extinct have been caught by fishermen. Caelacanths, thought to be extinct for millions of years, were caught off the coast of Madagascar in the 20th century. The deep-sea fish has jointed appendages alongside its fins that move in alternating rhythms, as though walking. It is popularly believed that this was one of the first fish to walk out of the sea on to land (Weinberg, 2000).

Fluctuations in temperature occurred over the last 3 million years, but overall a cooling rate has predominated, including several ice-ages (Shackleton, 1995). Fossil evidence shows that bacteria were the lone inhabitants of earth for millions of years, followed by a gradual evolution of other life forms (Fortey, 1997). However, before the Cambrian explosion, around 545 million years ago, there were extreme changes in climate and atmospheric conditions There are more fossils available from the Cambrian period, and the fossils are larger due to an increase in oxygen and mineral levels which led to an increase in nutrient availability. Another dramatic climate change occurred around 300 million years ago with an increase to over 30% oxygen in the atmosphere and a decrease in carbon dioxide. Again, the fossils reveal enlarged organisms. Insects with broader chest areas and larger wing spans make their appearance, and marine life shows a marked increase in size (Vrba, 2004). The increasing amount of oxygen in the atmosphere encouraged the migration of vertebrates from sea to land because of enlarged lung capacity (Graham, Dudley, Aguilar, & Gans, 1995).

During this period, new and larger life forms appeared around the world (Vrba, 1995a, 1995b) which may have included hominid ancestors (Asfaw et al., 1999). Dense vegetation flourished giving way to more grass savannahs and semi-arid conditions which is associated with bipedal development (Vrba, 2004). There is still a great deal of debate in the scientific world on the branch that humans took in mammalian evolution. However, the most popularly accepted hypothesis is that we are descended most recently from primates and hominids.

Animal Digestive Systems

A microbial ecosystem is necessary for digestion. For nutrition and energy needs, most organisms have similar internal systems to digest organic compounds, which are provided by other life forms. Bacteria are not limited to organic sources, but can also use mineral sources for nutrient and energy needs (Postgate, 1992). Cows and sheep do not have enzymes that can help them digest plant cellulose. However, each of these animals have large resident bacterial communities, e.g. *Ruminococcus* and *Fibrobactor,* in a special reservoir called the rumen. There, agile bacteria break down the cellulose to smaller products that the animals can digest. When eaten, the animals provide predigested nutrients (Singleton, 1999).

The digestive tract in each vertebrate organism is an enclosed tube from mouth to anus. The contents of the digestive tract are outside the body proper, which is protected from harm by the gastrointestinal lining. The epithelial cells lining the gut are a barrier of tightly intertwined junctions that allow nutrients and water to cross, but can prevent larger or harmful molecules from entering the blood or lymph systems (Whitney, Hamilton, & Rolfes, 1990).

The gastrointestinal tract provides a pathway for nutrients to be absorbed for the upkeep, energy and repair needs of the body. In the mouth the food is surrounded by saliva and enzymes, and becomes a bolus to be swallowed by the epiglottis. Reaching the stomach, the food is ground by gastric juices into chyme, which then passes into the small intestines. Here, the liver adds bile, via the gall bladder, to emulsify fats. The pancreas adds juice and enzymes to further degrade the carbohydrates, protein and fats. In the folds of the small intestine, villa pass the small nutrient molecules into the lymph and blood transport systems to be taken to the cells of the internal organs. Undigested foods, mainly fibers and cellulose are passed into the large intestine or colon, and what is not digested by the resident bacteria is passed out as wastes (Whitney, Hamilton, & Rolfes, 1990).

About 400 different strains of prokaryote bacterial communities inhabit the digestive tract. There are 10 times more bacteria in the human gut than there are other cells in the rest of the body. Most digestive bacteria are located in the large intestine (Bengmark, 1998). Communication is established between the epithelial cells and the gut bacteria in order to allow nutrients to cross into the bloodstream and to keep out dangerous toxins (Neish, 2002).

Shells

Thermatoga maritima, one of the bacterium that lives at an ocean vent has a protein outer covering that surrounds its membrane. Koneman (2002) describes the coverings as resembling the togas of ancient Rome, and goes on to say that is why Dr. Stetter named them *"Thermatoga* -the heat-loving, toga-wearing prokaryote that lives in the sea (*maritima*)"* (p.6). A salty environment draws the water out of bacterial cells, and if the bacteria cannot adapt to changing circumstances, they collapse and die (Singleton, 1999). Did the bacteria that first emerged from below the ocean vents' surface build protein cell walls around their outer membrane to protect themselves from the destructive ocean salt water? Were the protein cell walls the first shelters the bacteria were to build on the first step of evolution?

Minerals or Hard substances produced by living organisms

To ensure continuity throughout the eons of time that have passed since bacteria first appeared on this planet, organisms have developed outer coverings as a survival mechanism to protect the inner workings of the cell, and eventually to protect the tissues and organs of an evolved body. However, throughout the evolution of each organism, bacteria have always maintained access for either positive or negative contributions. The health of the organism is dependent on bacteria. However, in cases of disease or weakness, termination by the bacteria may be appropriate. Similar membrane structure allows the bacteria to enter the cell in cases of pathogenicity (Singleton, 1999).

All five kingdoms of life, Monera, Protoctista, Fungi, Plantae, and Animalia, use or produce minerals in some shape or form as a means of support or protection. Margulis and

Sagan (1995) outline six minerals, calcium, silicon, iron, manganese, barium, strontium, and the roles they play by members of the five kingdoms. They further show how, in animals, outer coverings of shells contain calcium carbonate, and the inner structure of bones and teeth contain calcium phosphate.

As each organism has evolved, so has the outer protective shell evolved according to diet and survival requirements of the organism. The majority of bacteria have thick outer protective walls that consist of lipids or carbohydrates. An outer capsule coating can be formed as a protection against infection or toxicity. The capsule can store nutrients or provide adhesion factors. In disease, a pathogen creates a capsule coating to escape detection by the immune system (Singleton, 1999).

In the endo-soymbiotic relationship between cyanobacteria and fungi, an essential point is that the cyanobacteria can survive without the fungus but, like all other life forms, the fungus cannot survive without the bacteria (Hudler, 1998). A conclusion could be made that lichen is one of the first land dwelling bacteria that formed a necessary outer shell, ascofungi, to survive in a new environment.

To survive on land, plants formed an outer cellulose (formed from bacteria and algae) structure that could hold water. To sustain growth in dry conditions, cellulose with lignin formed a stronger, outer coat of wood (Margulis & Sagan, 1995).

Varied marine life and soft-centered land animals create hard-backed shells. Fish and reptiles have an outer layer of scales. Some animals, including humans, have an outer covering of skin that contain sensory receptors with which to perceive the conditions of the environment in order to made aware of pleasurable sensations or danger. The sensation of pain is an immediate warning of danger. A skin covering keeps body parts inside so they can carry out their assigned tasks, and, at the same time, keeps out most harmful substances. Hair, fur, or feathers on skin not only regulate heat evaporation, but are also used to attract partners in order to perpetuate the species (Davis & Solomon, 1986).

The outer shell of skin, the epidermis which contains defensive bacteria, consists of several layers. In the lowest layer, cells divide and the new cells push up and replace the outer layer of dead cells which are sloughed off. As they progress towards the outer layer, the new cells produce keratin, which although it makes skin flexible, also makes it impermeable keeping unwanted intruders out. The inner layer of skin, the dermis, is made from collagen, and contains blood vessels. Below the dermis is a layer of insulating, fat-containing tissues (Davis & Solomon, 1986).

Within the protective covering of skin, humans and some animals possess a skeleton framework. The skeleton is a living, growing component of the internal structure, and while it supports the body, it is an anchor for muscles. For other creatures, their skeleton is a non-living part of the outer structure that holds the inner parts of the body together, and at the same time protects from assault and danger. Sea shells, beetle casings, and turtle shells are prime examples of outer skeleton shell coverings. The outer shell has two layers. The lower layer, the cuticle, consists of chitin and protein. The outer layer, the epicuticle, contains oils which help to conserve water. Before ecdysis, the organism seeks shelter in order to protect the exposed and vulnerable inner body from attack by predators. After ecdysis, a new chitin layer is produced which hardens, followed by a covering of protein and oils. The creature with its new protective covering is now ready to emerge from its temporary shelter (Davis & Solomon, 1986).

Around August 1, 2005, television world news, with pictures, reported that a rare blue lobster had been caught off the coast of Maine. It was later found that it was a lobster that had emerged from its old shell. The now famous lobster appeared to be encased in a somewhat hardened covering. It appears that the blue coloring was intended for the partially exposed lobster to blend in with the surroundings while it was constructing its new protective covering.

Margulis and Schwartz (1988) give interesting accounts of protective shells formed by soft bodied creatures. A remarkable fact is seen in the actinopoda, a sea-dwelling protoctist.

Its radial skeleton of crystalline sodium sulphate protrudes from the center at angles that match the longitude and latitude demarcations of the earth. Annelid worms have chitin bristles on their body surfaces that can be used for movement or anchorage. Ectoprocta, water-dwellers, may have just an outer layer of chitin, or skeleton made of calcium carbonate covered with a layer of chitin. Gastrotriches worms have covering or "a thin cuticle of lipoprotein and nonchitinous polysaccharide" (p.194). Arthropods and Onychophora have smooth coverings of chitin. Mollusks have shells made of calcium carbonate secreted by the sea creature, and they also have a strap of chitin with which they acquire food. Phorinida reside in tubes of chitin which they fortify with 'secretions impregnated with calcareous matter or strengthened with sand or shell fragments" (p. 212). The shell coverings which each animal creates according to its needs, or according to its environmental resources, helps preservation until reproduction, maybe several times over, has been accomplished.

Fetus to Newborn

Although trillions of bacteria dwell in and on the bodies of all humans, a fetus in the womb has no resident bacteria. Nutrients and oxygen are provided by the mother's blood via the umbilical cord and placenta which filter out bacteria. Heavey and Rowland (1999), in a review article, show how during birth, the baby acquires bacteria from the birth canal, and also from the immediate surroundings. How and what the baby is fed, is shown to affect the types and amount of bacteria that will colonize the gut of the growing infant.

Within a week of birth, an infant will normally obtain all the bacteria necessary for digestion and survival. Before teething, the oral cavity of an infant has only epithelial surfaces so the bacteria present colonize cheek and tongue areas. After teeth have erupted, anaerobic bacteria colonize the areas between the teeth and gums, and eventually most of the many species of microflora in the mouth will be facultative or obligate anaerobic bacteria (Marsh, 2000). Through enzymatic action, cells are able to release any oxygen that enters, preventing damage to the interior of the microbial cells (Carlsson, 2000).

The digestive tract of the newly born infant is prepared for the arrival of the bacterial communities necessary for survival. It appears that the infant gut is first colonized by aerobic bacteria which consume the oxygen allowing anaerobic bacteria to be established (Stark and Lee, 1982). Further, the bacteria that can reproduce the fastest become the dominant species (Falk, Hooper, Midvet, & Gordon, 1998). The first bacterial colonizers provide a niche for themselves that helps prevent later bacterial arrivals from setting up residency, and also the first arrivals affect the actions of genes in the gut lining (Hooper, et al, 2001).

Gradually more, and different, species of bacteria will be added to the digestive system as the child grows into adolescence, and it may not be until a few years later that resident bacteria will resemble those of an adult (Heavey and Rowland, 1999). However, each adult will have a different combination of bacteria than everyone else, which will remain mostly the same throughout life (Simon and Gorbach, 1984: Moore and Moore, 1995), but bacterial strains may change according to environmental factors (Singleton, 1995).

Mouth

In the animal kingdom, the oral cavity is the point of entry for nourishment. The food is chewed, and enzymes in the saliva break down the food until it can be swallowed. However, the mouth is not only for chewing food. Editors Kuramitzu & Ellen (2000) have compiled the works of several scientists which describe the role of bacteria in the oral cavity of humans. The mouth, helped by the lips, is an important bacterial line of defense in preventing invading pathogens and viruses from entering the digestive system. Defensive, microbial communities inhabit as many areas as possible on the lips, tongue, gums, and cheeks to prevent viral microbes from getting established (Marsh, 2000).

Another barrier against invading pathogens occurs when several bacterial strains build plaque that attaches to the teeth. Other bacteria form a biofilm which adheres to the plaque, preventing the protective bacteria from being swallowed. Pathogens, repelled by the resident microflora, are unable to gain a foothold in the mouth (Marsh, 2000).

Most of the bacteria in the mouth are anaerobic, some can grow with or without oxygen or they require reduced amounts of oxygen. Those who cannot tolerate any oxygen associate closely with aerobic bacteria (Carlsson, 2000). Postgate (1992) posits that everyone develops their own kinds of bacteria within their nose and larynx. The bacterial strains develop early and repel everyone else's personal bacteria.

Oesophagus

Some animals have been found to have bacteria residing on the skin surfaces of their esophagus. However, any human bacteria in the esophagus do not adhere to the skin surfaces, and are only transitory (Tannock, 1995).

Stomach

Lactobacilli and streptococci bacteria, swallowed with food and saliva, pass quickly through the stomach by producing enzymes that increase the pH around the bacteria. It was thought that the highly acidic contents of the stomach prevented bacterial colonization. New discoveries show that *Helicobacter pylori* bacteria inhabit the stomachs of even healthy individuals. There are pockets in the epithelial lining of the stomach which have a neutral pH and this is where the *H. pylori,* after covering themselves with an alkaline coating, congregate (Wassener, 2005).

Food is broken into smaller particles by stomach acids, and the fats are broken apart by bile. Then, although passing through quickly, nutrient absorption occurs in the duodenum and jejunam. Food progresses more slowly through the ileum, and it is here that increasing communities of obligate anaerobes are located (Tannock, 1995). Bacteria are responsible for breaking down the food to even smaller molecules for absorption into the intestinal lining (Wassener, 2005).

Colon

The colon, or large bowel, where content motility slows considerably, is where water and fluids are absorbed from the lumen. Home to a great number of microbe communities, it is in the colon that the bacteria are responsible for extracting nutrients for themselves from undigested proteins, carbohydrates, fats, and cellulose. The bacteria are also responsible for breaking down and removing wastes. Proteins that enter the colon are degraded by enzymes into peptides, then into amino acids. Any ammonia produced is used by the bacteria for needed proteins (Tannock, 1995). A striking feature is that all the bacteria in the gastrointestinal tract are either obligate anaerobe or facultative anaerobe prokaryotes, the bacteria that originated over 4.5 billion years. There are ten obligate anaerobe bacteria for every facultative anaerobe bacterium in the colon (Simon & Gorbach, 1984).

Vitamins that the body can't produce, but are essential for organisms to function correctly, are synthesized or broken down by the microflora communities within the colon. Vitamin A is synthesized by bacteria, and also the B vitamins, biotin, pantothenic acid, pyroxidine, and riboflavin (Postgate, 1992; Tannock 1995). Cobalt necessary for blood formation is contained in another vitamin, B12, and is also synthesized by bacteria. Vitamin K, also synthesized by the microflora, helps blood to clot. *Escherichia coli* are involved in synthesizing all these vitamins (Tannock, 1995).

About half of leftover contents within the colon, or large bowel, are bacterial cells. Energy and nutrients for the microbes are subtracted from undigested foods which supply undigested fermentable carbohydrates to resident bacteria, nourishing them so that they can multiply, and also help replace the microbes lost in the evacuation of wastes. If any oxygen manages to invade the area, it is metabolically reduced by the resident bacteria (Tannock, 1995).

The indigenous microbiota in the intestines can repel pathogenic strains of bacteria by preventing their attachment to the gut lining, by blocking the invasion of the gut lining, and by taking up all the available nutrients. The unwelcome pathogens are passed out in faeces. To prevent over-crowding, to keep a balance of healthy bacteria in the gastro-intestinal tract, and to help prevent shortages of nutrients or energy, each bacterium is equipped with a mechanism for self destruction (Postgate, 1992).

Bacterial Communication

Communication between bacteria is carried out through signal induction and quorum sensing. To monitor environmental conditions, bacteria communicate within strains, between strains, and with different strains of bacteria (Singleton 1999). Information about the environment comes through the exchange of solutes during collisions with other molecules (signal induction). Nutrients pass into the cell, wastes pass out with the aid of solutes. In charge of communications of outside conditions is solute K which operates from the surface of the cell, and transmits messages into the cell, where chemicals in the outside environment are detected. Interaction occurs between cell proteins, and there are changes in gene expression. Through signal induction, the diverse microbiota avoid competition for food or energy, and are able to grow, reproduce, and protect each other. Exchanging information leads to adaptive responses and enables diverse bacterial communities to live safely together in enclosed environments (Carlsson, 2000).

Quorum Sensing in bacterial communication shows how bacteria signal to each other to congregate in order to achieve a common objective, e.g. luminescence in certain fish to repel predators. Using quorum sensing, the number of bacteria in the area can be calculated. When enough bacteria have come together (a quorum) the bacterial signaling secretions (autoinducers) activate the lux genes, and light is produced. Different autoinducers activate different genes in order to produce different effects. Sometimes, luminescence is required to alert others to food, to signal danger, or to attract mates. Other times, toxic substance can be formulated to repel or destroy predators (Singleton, 1999).

Bacteriome

A bacteriome, found next to the ovary of rice weevil, and maternally inherited, contain bacteryocytes within which reside *E. coli,* bacteria that are almost identical with bacteria found in the colon. The bacteria are instrumental in synthesizing B vitamins. Without the vitamins, the weevils are stunted, cannot fly, and their reproductive mechanisms are compromised (Margulis & Sagan, 2002). A lack of indigenous microflora leads to a weakened immune system, lowered vitamin count, and malfunction of critical organs. Also, the assimilation of nutrients is compromised (Darling, 2000).

Antibiotics

There has been a proliferation of antibiotic medications in the last few decades, leading to the disturbances and breakdown of bacterial communities within the human body. Many antibiotics attack and destroy infectious bacteria that do not have cell walls. Unfortunately, the natural, protective microflora become decimated because many of the anaerobe bacteria in the digestive system also have no outer cell walls (Tannock, 1995). Other bacteria are antibiotic resistant. The plasmid content of the bacterial pathogen *Yersinia pestis* makes it especially resistant to antibiotics (Carniel, 2002).

Antibiotics were introduced to kill pathogenic bacteria. However, antibiotics are site specific, and some can only be effective for a few bacteria, whereas others can be effective for a wider range of bacteria. Some antibiotics kill the pathogenic bacteria while others stop the bacterial action. Action of the antibiotic may be on the cell wall, the inner membrane, or in the reproductive mechanisms of the DNA. Few diseases are cured or controlled by antibiotics. Many antibiotics work at different stages of the cells' formation and growth. To act effectively, two antibiotics working together are sometimes required (Singleton, 1999).

There are strains of bacteria that can produce enzymes to effectively block or inactivate antibiotic action. Others do not have the target site, e.g. wall-less bacteria are not affected by antibiotics that degrade cell walls. Yet other bacteria can block antibiotics from reaching their target sites. Some antibiotics interfere with certain synthesizing actions within the cell, but bacteria differ in the synthesis of their requirements, and many do not carry out the action required by the antibiotic for the antibiotic to work. Since the introduction of antibiotic treatment, many bacteria have evolved to successfully counter attack and negate antibiotic action. New antibiotics are continually being invented only to have the bacteria devise retaliatory actions in return (Singleton, 1999).

Continuous antibiotic use can deplete the body's healthy bacteria and allow the entrance, adherence, and proliferation of pathogenic bacteria. After antibiotic treatment, which also suppresses the normally protective bacteria, it has been found that *Clostridium difficile*, usually benign in childhood (Neish, 2000), proliferates in the colon and then gains access to the bloodstream by secreting two toxins that act together to penetrate the lining of the digestive tract. Antibiotics have been used in cancer therapy to suppress pathogenic bacteria. Sadly, suppression has led to a decrease in normal healthy bacteria, leading to problems for patients with suppressed immune systems (Tannock, 1995). Another problem, as Darling (2000) points out, is that antibiotics destroy *L. acidophilis* and other resident microflora. In the ensuing empty space, fungi, which antibiotics do not affect, take over and cause many intestinal problems.

Indiscriminate use of antibiotics is ill advised. It should be known, before administration, what action the antibiotic will take within the body, and will its use be effective. Long term use can be disastrous because antibiotics can decimate the vitally necessary healthy microflora in the gastrointestinal tract, allowing antibiotic resistant pathogenic bacteria to become established, which can lead to serious digestive and excretory illnesses, and in some cases death (Singleton, 1999).

In October 2004, a close friend of the author underwent surgery for the removal of a section of intestine due to diverticulitis, where the walls of the colon weaken, and small cavities appear in which wastes accumulate. The wastes become stagnant causing fever and pain. For many years, under medical supervision, antibiotics had been taken to treat the condition. After surgery, the bacterium *Clostridium difficile,* which secretes toxins A and B (Tannock, 1995), proliferated and due to long-term antibiotic use, there were insufficient resident healthy bacteria to prevent the fatal take-over by *C. difficile.* Within a week, blood clots, heart, and lung problems developed. She was put on a respirator. With no chance of survival, she was taken off the respirator, and died within 15 minutes.

Probiotics

Sankaran (2000) describes a probiotic as a "viable lactic acid bacteria" (p. 202) cultured in a food supplement, e.g. yogurt. Also, to be effective if administered, the probiotic bacteria should be the same strain of bacteria that already inhabits the host digestive tract. Further, the probiotic bacteria must be acid resistant so that they survive the passage through the stomach, and they must also be able to attach to the gut lining, and when there, to proliferate. The author also states that health benefits of probiotics include relief for lactose intolerant people, better digestion, and reduced growth of pathogenic bacteria. The case is also made that the age old traditions of eating yogurt and cheese at the conclusion of meals provides the same ingredients that probiotics provide.

Lactobacillus acidophilus. as its second name implies, can survive in an acidic environment so, eaten in yogurt, it passes through the stomach and then inhabits the intestinal lining, its wastes driving off pathogenic bacteria (Darling, 2000). Lactobacilli, only a small component of intestinal bacteria, came into favor because of their safety in long time use. The greater numbers of bifidobacteria inhabiting the gastro-intestinal tract have led to an increase in their use in probiotic measures. Nutrients promoted as probiotic, are now regulated by the same standards required of pharmaceuticals (Tannock, 1995).

Lactobacillus bulgaris and *Streptococcus thermophilus* are two bacterial strains used together in starter cultures for yogurt by fermenting milk sugars into lactic acid. The two bacteria together accelerate the rate of curdling. *L. bulgaris* prevents the proliferation of *Salmonella* pathogens in milk. *Lactobacillus acidophilus* is used in starter cultures to make acidophilus milk. The cultured milk is effective because it is a health benefit for the intestinal tract, and degrades some strains of pathogenic bacteria (Sankaran, 2000).

Evolution of Species Theories

Evolution, in the biological sciences, explains how a simple organism (e.g. a bacterium) changes, over eons of time, into a complex organism (e.g. a human being) via the actions of genes residing on the chromosomes, and also the effects of the environment on the organism. Complex humans have only a few thousand more genes than small roundworms (Balon, 2004). Margulis and Sagan (2002) explain how an organism's genes carry out their more complex activities through gene duplication. The copies of the genes can be altered to fulfill the needs of evolution and still leave the originals intact ready for further, different duplications. Most scientists agree that all species are descended from previous species, but problems arise over deciding how evolution took the path it did.

Considering the complexity of evolution, there are naturally, many theories of how organisms evolved. The evolutionary theories of Charles Darwin (1859), natural selection and survival of the fittest, have been the most accepted theories by the majority of the scientific community, and for many years have been taught in most schools and colleges.

Natural selection is the theory of evolution in which genes mutate at random. If the resultant organism has a survival advantage over other individuals of the same species in the same environment, that organism tends to be replicated and become a dominant factor in the evolution of the species (Darwin, 1859). However, the majority of mutations are lethal and so only few mutations survive (Postgate, 1992).

Survival of the fittest is the ability of an evolved organism to grow to maturity and successfully reproduce healthy offspring (Darwin, 1859). Mayr (2001), in the light of over 140 years of advanced technology, explains and further supports Darwinian theories. In reply to the query of why species exist, Mayr (2001) posits that hybrids, "genetic backcrosses" (p.169), are not usually viable or cannot reproduce. Therefore, members of the same species have to mate to produce viable and fertile offspring. However, Margulis and Sagan (2002) point out that no new species are created by natural selection, just a more survivable member of the same species.

Another theory of evolution has been proposed by a growing body of scientists from a re-emerging field of Evolutionary Developmental Biology. These scientists examine new developmental mechanisms that affect physical and behavioral changes in an embryo, or maybe even later into its adult life, that will ultimately affect the evolution of its species. New developmental pathways within the embryo are carried out by hormonal signaling. Hall (2004) emphasizes that it is not the signals passing between the cells within the embryo that causes new pathways to form, but by the signals that are coming from individuals in the external environment. In the separate studies of Dodson, Stearns, Hall, & Gilbert (as cited in Table 1.1, Hall, 2004), it was found that the individuals could be from the same or a different species. In effect, the external environment is guiding the formation and growth of an embryo that has a better chance of survival and reproduction when it enters the outside world.

The required alterations in the structure of the embryo could come from the outside environment through fluctuations in heat, light, or temperature (Youson, 2004). Changes can also be effected by diet, the seasons, population density (Larsen, 2004), raising or lowering of salinity, or extreme stress (Reid, 2004). Organisms of the same (Abouheif, 2004) or different species (Hall, 2004) surrounding the adults carrying the embryos can also affect embryonic changes. Predators have effected changes in embryos of prey where no changes have taken place when the predators were not present (Hall, 2004). Low environmental temperatures in certain reptile species result in males, and higher environmental temperatures result in females (Abouheif, 2004). In Evolutionary Developmental Biology, the premise is that the outside environment dictates to the growing embryo what is needed to survive in the prevailing conditions. A big advantage is that many organisms are affected at one time, and so have a greater chance of survival (Reid, 2004).

In their book, *Acquiring Genomes: A Theory of the Origins of Species,* Margulis and Sagan (2002) suggest another theory of evolution. The two scientists posit that when two or more "differently named organisms" (p. 7) merge, the symbiogenesis provided by the combined genes of both species leads to greater and more advantageous genetic variability, and can result in new species. Further, gene pools are merged by organisms that coexist and provide benefits to each other. An example of merged gene pools is when one organism becomes the food of the other and the ingested organism becomes dependent on the waste products of the predator. Members of the same species may have problems if their bacterial populations are different, missing, or diminished. Offspring may not grow as well or be productive. In some cases, the ingestion of bacteria can lead to new species. Acquisition or re-pairing of DNA sequences, crossing over of chromosomes, gene copying, or rearrangement affect the acquisition of new genetic material. Margulis and Sagan (2002) mention one surprising result, "When the complete sequencing of the human genome was announced...some 250 of the more than 30,000 human genes of our bodies have come directly from bacteria" (p.76). It is also suggested that the transmission of genes from bacteria to humans may be accomplished by viruses. Micorobes are credited with being the "engines of evolutionary change" (p.87). Further the two authors succinctly express the fact,

> That bacteria duplicate, transfer, digest, and in other ways lose and gain genes is an essential aspect of the evolutionary saga. The speed, volume, and antiquity of bacterial gene-trading activities underlie the evolution of all the rest of life on Earth. (p.85)

The ability to carry out extremely complicated maneuvers in an efficient manner enabled bacteria to be the true ancestors of all life's organisms.

A further step describes that, when stressed, bacteria joined forces with each other to evolve into protoctists that developed strategies for survival and evolution in an environment of increasing oxygen. However, the authors, and a colleague of theirs, David Searcy, (as cited in Margulis & Sagan, 2000) say that the bacteria "still 'remember' their origins because the ancient environment is built into the metabolism" (p.153).

Instead of the guesswork involved in random mutation, Margulis and Sagan (2002) maintain that symbiogenesis provides for a higher volume and success rate in the evolution of species, whereas in random mutation, few organisms are affected in a healthy way, and most face extinction before their genes are passed on to their descendants. It is also proposed that although random mutation can improve a member of a species, it cannot cross the species barrier. However, viable inherited characteristics come together when two gene pools merge, and so can cross species barriers and effect variety and dramatic change. The resultant changes are irreversible. The sudden dramatic changes due to merged genomes are borne out by the fossil record where new species occur in discrete jumps, known as Punctuated Equilibrium.

Brace (1998) presents several theories and ideas of human evolution. Clades portrays the existence of the same ancestral line with the divergence of organisms to a different form. Orthogenesis refers to "unknown or supernatural guidance" (p.58), and evolution can only proceed in one direction. Genetic Drift occurs where there is a catastrophic event and only survivors pass on their genes. Preadaptation suggests that something that is already formed in the organism is changed as an adaptation to the environment, e.g. fins became limbs. Neoteny is a supposed retention of something that stays the same from infancy to adulthood, e.g. large head and small face. Sexual dimorphism explores how women developed as child-bearing followed by providing milk for the infants, and men developed as protectors and hunters. Human Adaptation is concerned with the physical or cultural environment within which the organism grows.

Bacteria and Intelligence

Bacteria must have a form of brain or a base of intelligence. They can alter their routes and destinations by checking, among other things, nutrient, oxygen, saline, and temperature levels. They must also have memory capabilities to be able to identify and compare environmental changes from one moment to the next (Snedden, 2000). In other words, if the definition of intelligence is the ability to survive in one's environment (dissertation author's definition), the ability of bacteria to survive 4.5 billion years makes them top candidates for that definition.

Evolution and Nerve Development

Simple animal forms, e.g. cnidarians and planarians, as outlined by Davis and Solomon (1986), have one opening to the outside world that takes in nutrients and also passes out wastes. The two authors also show how further along the evolutionary timeline, worms, e.g. Phylum Nemertinea, have the first separate enclosed digestive systems, "a tube- within-a-tube body plan" (p. 316) having an entrance (mouth) for nutrients and an exit (anus) for waste products.

The hydra (a cnidarian polyp) is credited with having the first network of nerves that are spread out within the whole organism, but has no central brain. However, a flatworm is shown as having branched nerves along the length of its body, with a projection of the nervous system into the head of the animal. The nervous system in the head is divided into two bundles of ganglia with two cords connecting them to the nerves along the worm's body. The projected assembly of nerves in the front end of the flatworm gives it a distinct advantage for quickly locating food or avoiding danger (Davis & Solomon, 1986). This raises an interesting question. Are the two separate bundles of ganglia in the head of the flatworm the forerunners of the two hemispheres in the human brain?

The rediscovery of the enteric nervous system and its importance in the health and welfare of the body has created a new area of research in the medical field. In his book, *The 2nd Brain: Your Gut Has a Mind of Its Own* (1998), Dr. Gershon, writes of his efforts to re-convince the medical community that there is an independent, self regulating nervous system within the gastrointestinal tract. There was a great deal of opposition to his ideas, and many scientists tried to prove that his ideas were inconsistent with the actual physiology of the nervous system. In support of his proposal, Gershon recounts how Auerbach, in the 1860's, first demonstrated that the gastrointestinal system contains nerve cells and their corresponding fibers. He also relates how Bayliss and Starling (1899) after cutting off communication between the gut and the cranial brain of a living dog, discovered that food would still continue to move down the intestine when sufficient pressure was applied. Bayliss and Starling (1899) wrote,

> The peristaltic contractions are true co-ordinated reflexes, started out by the local
>
> nervous mechanism (Auerbach's plexus). They are independent of the connection
>
> of the gut with the central nervous system. They travel only in one direction, from
>
> above downward and they are abolished on paralyzing the local nervous apparatus
>
> by means of nicotine or cocaine. (March 17, 1899. pp.142-143)

Dr. Gershon (1998) also describes Ulrich Trendelendberg's experiment in 1917 using the intestine of a guinea pig that had been completely removed from the animal and placed in a warm bath containing nutrients and oxygen. When pressure was applied to the severed intestine, Trendelenberg also noticed a downward movement in the intestinal contents and named the muscle movements "peristaltic reflex" (Gershon, 1998, p. 7). There is no other organ in the body that can function without input from the brain. If any other organ is severed from the nervous system, it dies (Trendelenberg, 1917, as cited in Gershon, 1999). Although the scientific discoveries were published and recognized, they became lost due to human machinations and new discoveries.

A colorful description is given by Dr. Gershon (1998) of medical politics early in the 20th century. In 1921, J. N. Langley, editor and owner of the Journal of Physiology, published his magnificent opus, *The Autonomic Nervous System,* which, according to Dr. Gershon, is still the cornerstone of medical textbooks describing the nervous systems of the body. Langley agreed with previous research that there are three divisions in the autonomic nervous system; the parasympathetic, the sympathetic, and the enteric nervous system and he outlined all three in his book. Unfortunately, after Langley died, the Journal of Physiology was taken over by scientists who had been rebuffed or harshly criticized by Langley. The new owners of the journal concentrated on the emerging science of neurotransmitters that enervated the parasympathetic and sympathetic nervous systems, bringing the two systems into prominence. The enteric nervous system was relegated to being a part of the of the parasympathetic nervous system.

However, as Bayliss and Starling (1899), along with further research by Trendelenberg (1917), demonstrated, it is the enteric nervous system that directs the digestion of food and the breakdown of wastes without any input from the cranial brain. The vagus nerve, the nerve that connects the cranial brain to the digestive tract, can pass on the message from the cranial brain if defensive action is needed e.g. to initiate vomiting in a case of poisoning, but it is the enteric nervous system that directs and carries out the needed action. It has been shown that the digestive tract can still carry out its tasks of digestion and evacuation even when the vagus nerve is severed. Also, if the colon is removed and placed in a life supporting solution, the downward passage of digestive matter still occurs (Gershon, 1998).

The personal research of Dr. Gershon (1998) found that the neurons in the cranial brain and the enteric nervous system produce the same neurotransmitters. Serotonin has been shown to be the neurotransmitter that initiates the peristaltic reflex in the gut without any input from the central or peripheral nervous systems. When serotonin secretion stops, so does the peristaltic reflex.. Also, serotonin is responsible for producing "95% of the body's serotonin in the gut" (Gershon, 1998, p. xii). Dr. Gershon also shows how the enteric nervous system sends out nerve connections to the gall bladder and pancreas thus controlling the actions of the two organs.

Chapter 3

Ancient Life

There are rocks in West Greenland that are 3.8 to 3.9 billion years old. Within these rocks is fossil evidence of ancient bacteria. The carbon within these fossils shows it to be the same carbon that is found in all earth's life forms. This legacy of ancient bacteria is thought to be the first evidence of life on earth (Harper, 2005). In the studies of Krumholz and McKinley; Pederson; Tseng, Onstatt, and Person (as cited in Monastersky, 1997) land related discoveries around the world have found bacteria that are 80 to160 million years old enclosed in sandstone, granite, igneous, or sedimentary rocks up to 3 kilometers underground. The thermophilic bacteria live in temperatures up to 75 degrees Centigrade, and obtain nutrients from organic or inorganic compounds that filter through the rocks. The limited food supply prevents growth and reproduction, and the bacteria appear to be in a quiescent state. More recent discoveries have found live bacteria in rocks 3 1/2 kilometers deep in a South African gold mine. The bacteria are over 200 million years old. Nutrients are provided by methane and propane gas, but in such limited amounts that only enough energy is obtained by the bacteria to divide and reproduce every 1,000 years (Harper, 2005).

In 1991, scientists aboard Alvin, a submersible craft that can probe the ocean depths, observed bacteria as the first colonists in newly erupted vents on the ocean floor. They saw bacteria emerge from below the earth's crust, and flow through the thermal vents in huge, white, billowing clouds. The clouds of bacteria formed large white mats on the ocean floor. The researchers commented that continued observations would not be possible because if light in any form was introduced at the vents, it would disrupt the natural life patterns of the vent inhabitants. On a 1992 visit of Alvin to the same location, most of the bacterial mats had disappeared, and there was now marine life around the vents (Stover, 1995).

The temperature at the vents has been measured as high as 400 degrees C, and bacteria are the only organisms able to withstand such tremendous heat. However, one inch away from the vents, the temperature drops to 2 degrees C. So it is not heat but thermochemicals that provide the energy for creatures that later live around the vents in close to freezing temperatures. Bacteria use hydrogen sulfide for their energy needs. The vent marine life ingest the bacteria, and so form a symbiotic relationship with the bacteria. When a vent shuts off and stops supplying thermochemicals, all life around that particular vent ceases (Stover, 1995).

Shells

A bacterium, when it emerges from the thermal vents on the ocean floor, produces a protein coat to protect itself from the drying out effects of the ocean waters' salt content (Singleton, 1999). During the 2004 excursion of Alvin, Pompeii worms were collected from the hot water vents on the ocean floor. The 4 inch worms, encased in thin tubes, were sticking out from the sides of the vents, the water at the base of the tubes measuring 81 degrees C. However, the heads of the worms were sticking out of the tube to where the water temperature was 22 degrees C. Hot water and chemicals were being sprayed over the worms but they seemed to be protected by a coat of bacteria appended to their backs with hair like structures. The bacteria were feeding on mucus exuding from the worm surface. It is thought that the worms feed on the bacteria (Roach, 2005).

The largest organisms around the lava flowing thermal vents are tube worms that can grow over 8 feet tall. Inside each tube worm are billions of bacteria, and having no mouth, digestive system, or anus, each worm is dependent on the bacteria for nutrition. Hemoglobin in the red-plumed tips of the tube worms transports hydrogen sulfide from the vent waters to the bacteria and provide their energy needs. In return the bacteria relay carbon compounds from the hydrogen sulfide which provide nutrients for the worms. Eggs and sperm from the tube worms come together in the ocean water, and new worms form which have mouths and digestive systems, and so are able to ingest bacteria (Cavanaugh, 1995). However, as the worms grow they lose their mouths and digestive systems and the bacteria, now enclosed in an organ called the trophosome, provide all their nutrient needs. The walls of the tube worms consist of chitin, forming a strong outer shell (Sea & Sky, 2005).

In an effort to see how tube worms colonize a thermal vent, lavae of the red-tipped tube worm Raftia pachyptila were collected and raised under the pressures of ocean depths. The lavae were found to have a 38 day supply of food and it is posited that because they were found floating in the upward gushing vent plumes, the food supply would last them until they found another thermal vent to inhabit and colonize (Marsh, Mullineux, Manahan, & Young, 2001).

Bacteria found in ocean thermal vents have a genetic link to earth's early bacteria. Two scientists explored a cave called Queva de la Luz in the Rain Forest in Southern Mexico that has an atmosphere of hydrogen sulfide, and walls that are heavily coated with sulfuric acid. Wearing masks to protect themselves from the deadly toxic fumes, the scientists pointed out the snotities, stalactite-like formations, hanging from the roof of the cave. The snotities consist of millions of single cell bacteria encased in a slimy substance. In the extremely acidic atmosphere, the bacteria provide themselves with protective shells of the slime material. Phlegm balls of bacteria, also encased in a slimy substance, lay in the floor waters of the cave. Further exploration of the cave revealed many communities of bacteria all feeding on the chemical energy hydrogen sulfide. The narrator of the show, Neil de Grasse Tyson, posited that, due to the bombardment of meteorites and comets on the surface of the early earth, maybe the bacteria sought the safety of the caves below ground (Harper, 2005).

Stromatolites, huge mushroom shaped rocks standing in shallow ocean waters, were built, layer upon layer, by photosynthetic cyanobacteria. Apparently, only the surface of the rocks are covered by live cyanobacteria which exude a sticky substance that collects sand and debris, and forms a shield to protect the cyanobacteria from the sun. Each layer of a stromatolite is about 1/2 mm. thick, and is repeated over and over again as the cyanobacteria radiate up to the surface, and another layer of dust and sand adheres to the sticky substance produced by the migrating cyanobacteria. In the hills of Western Australia, an area that has remained unchanged for 3 1/2 billion years, fossils abound that show bacteria fossil configurations of black mats or stromatolites (Harper, 2005).

An abundant and vastly important bacteria needed for the survival of life on planet Earth was recently discovered cleverly hiding out on the surface of the world's oceans by Sally Chisholm and colleagues. Named *Prochlorococcus,* by their discoverers, the one- celled cyanobacteria pack over 20,000 cells in one drop of water. Along with the other phytoplankton, e.g. diatoms and algae, the tiny creatures, through photosynthesis, provide about half of the world's oxygen and remove as much carbon dioxide. The scientists, who have identified over 35 strains of *Prochlorcoccus* were amazed that anything so important to the survival of life could have been overlooked for so long (Chisholm, 2003).

Fungi & Plants

In natural habitats there are few of the plant diseases that occur in regulated, planted crops, and a vast amount of research has shown that a healthy plant is the result of a healthy root system. Mycorrhiza, the symbiotic relationship between fungus and bacteria that surrounds the plant's roots, has been shown to be the agent for root health. As it surrounds the roots, mycorrhiza improves "plant growth, tolerance to soil toxicity (heavy metals, salinity), transplant success, crop uniformity, root development, soil drought tolerance, disease tolerance" (Linderman, 2005, pp. 9-10). Environmental factors such as temperature and the health of the surrounding soil, also the nutrition supplying capabilities of the plant and its age, affect the robustness, or lack of it, of the mychorriza. Compost is recommended as a beneficial adjunct to soil preparation. The many different strains of bacteria are essential partners in promoting the well-being of the plant (Linderman, 2005).

It has been found that bacteria not only promote health, but can act in an antagonistic manner to soil pathogens. A new *Paenibacillus* strain of bacteria was discovered in the research of Budi, van Tuinan, Martinotti, & Gianinazzi (1999). *Paenibacillus* was found to prevent the invasion of fungal pathogens, to curtail the proliferation of pathogens, and at the same time was instrumental in promoting healthy growth of sorghum plants. Mamatha, Bagyaraj, & Jaganath (2002) showed how adding the bacterium *Bacillus coagulans* to fungi and therefore forming mycorrhiza, on plant root systems, also prevented pathogenic invasion while assisting in the growth of mulberry and papaya plants.

One of the main functions of mycorrhiza is to procure nutrients from the surrounding soil (Linderman, 2005). However, fungi cannot transfer the long-chained molecules of humic acid, found in decayed compost material, into the plant cells. According to an article by Julien, D. (2000), Dr. Robert Linderman has shown that the fungi create pathways to the cells of the plant, through which bacteria carry long-chained molecules of humic acid up to the fungal arbuscule. The plant is then able to absorb the arbuscle and so obtain nutrients. The improved growth of plants by using compost was thought to be due to the moisture holding properties of the compost. Now, the ability of bacteria to carry the large molecules of humic acid into the plant cells explains the reason for plants to excel when grown in compost. Also, because bacteria and fungus transfer phosphorus into the plant, it is unnecessary to add additional phosphorus to the soil. Added phosphorus run-off from gardens and lawns creates toxic algae proliferation on the surface of rivers and streams leading to the decimation of natural inhabitants.

Bacteriome

Arthropods (invertebrate insects) have been found to house anaerobic, symbiont bacteria. Nutrients are provided to the arthropods by the bacteria, especially in insects that feed on only one food, e.g. blood, tree sap, etc. Nitrogen wastes are also taken care of by the bacteria. The presence of bacteria in the insects helps to ensure a healthy reproductive system, a large egg production, and positive embryonic growth. B complex vitamins, necessary for the health, growth, and survival of the arthropods, are synthesized by the bacteria. Bacteria have evolved with the insects for over 5 to 8 million years (Akman & Askoy, 2001).

Eggs in the ovaries of homoptera insects, e.g. cicadas, psyllids, and scale insects, contain endosymbiont bacteria. The bacteria are contained in bacteriocytes enclosed in a bacteriome, which lies in the abdomen. Research on the 16S rRNA of pseudococcid bacteria, *Antonina crawii.* was investigated and three nucleotide sequences were found. Two sequences are similar sequences of *Proteobacterium,* and the third sequence is very similar to sequences of *Spiroplama* spp, found in ladybird beetles. Both *Proteobacteria* sequences are found in the bacteriome, showing vertical transmission. Very few sequences of *Spiroplasma* were in the bacteriome, but were found in other body tissues. Ladybird beetles, which contain *Spiroplasma* spp. ingest homoptera insects which may illustrate horizontal transmission (Fukatsu & Nicoh, 2000).

The individual studies of Brown, Tremblay, and Shigenoba (as cited in Normack 2004) show how a few armored scale insects are obligate chimeras. The insects have two genetically different sets of genes that operate in all their life stages. In meiosis, instead of the usual 2 haploid chromosomes of female egg and male sperm that becomes the embryo, a third cell that contains 3 copies of each chromosome merges with the embryo forming a pentaploid cell containing 5 copies of each chromosome. After multiplying, the cells form the bacteriome of the embryo. As the embryo grows, bacteria from the mother's bacteriome enter the embryo's bacteriome. The bacteria are speculated to help provide nutrients.

A further study by Majerus (as cited in Normack, 2004), demonstrates that in the bacteriome of the armored scale insects, bacteria have a controlling role. The bacteria alter the host's genes in order to destroy or feminize males for reasons of survival for the bacteria. Bacteria come from the maternal inheritance and so producing too many male progeny does not promote their survival. Also, the inheritance from female chromosomes provide identical females that help each other and do not engage in conflict to obtain necessary nutrients as would occur in insects with both male and female inheritances.

Male-killing Bacteria

The research of Hurst and Jiggens (2004), reveals that bacteria are extremely important in insects. The new area of research involves the bacteriome in female invertebrates. Housed in the bacteriome are male-killing bacteria that can affect the reproductive system of each female embryo. Transmission of male-killing bacteria can occur between two hosts, within or across species, or it can be accomplished through maternal inheritance.

Further, the higher number of females produced is accomplished by female embryos eating internally damaged male eggs, which leads to less sibling rivalry for nutrients, and also for less chance of inbreeding, a factor for negative consequences on progeny. Male-killing bacteria fulfill two roles in female hosts. They can protect and immunize against infection, and also increase production of progeny. Per cent ratios of female to male surviving populations depend on the types of hosts, and the requirements and conditions of the environment.

Host survival rates increase or decrease in relation to the number of male-killing bacteria in the bacteriome. If too few male embryos are killed, there is too much competition for nutrients, If too many females are produced, many may not mate and so decrease the population. Other bacteria in the bacteriome can effect the transmission of pathogenic bacteria to different organisms (Hurst & Jiggens, 2004).

Gastrointestinal Health Interactions between Bacteria and Humans

In the article, *Gut flora in health and disease* (2003), Guarner and Malagelada review scientific research showing how the bacterial communities within the large intestine of the digestive tract play an indispensable, interactive role in the health and physiology of the human organism. The resident microflora obtain nutrients and energy from undigested food products, assist immune functions, help the growth and spread of epithelial cells that line the gastrointestinal tract, and help protect the epithelial lining from the penetration of differing strains of bacteria that can become lethal after entering the bloodstream.

Bullen, Tearle, & Stewart's study (as cited in Heavey and Rowland, 1999) showed that breast milk passes on bacteria from the mother that help build an infant's intestinal microflora. Babies fed breast milk were once found to be more immune to gastro-intestinal upsets than babies fed formula. However, formulas are now manufactured that more closely replicate breast milk, but there are still some strains of bacteria missing from the microflora of formula fed babies.

For the development of healthy blood vessels in a newborn, it is essential for an interaction to take place between bacteria and the Paneth cells that reside in the intestinal lining of the digestive tract. Research shows that if no bacteria are present, blood vessels do not form. However, when colonies of bacteria are introduced, blood vessels start to develop. A further result shows that if the *Bacteroides tactaiotaomicron* strain of bacteria is introduced alone, blood vessel formation ensues. The pathways of the bacteria were found to be through the Paneth cells (Stoppenbeck, 2002).

Within the 400 strains of bacteria in the human colon there is a ratio of 10 to 1 of anaerobic versus aerobic bacteria, with dominant and sub-dominant strains (Moore & Moore, 1995; Simon & Gorbach, 1984). Bacteria are involved in transporting calcium, magnesium, and iron into the bloodstream. They are also involved in the production of short chain fatty acids which are necessary for a healthy blood supply, and for increasing energy levels. Bacteria also help produce short chain fatty acids that aid in promoting the growth, proliferation, and differentiation of epithelial cells (Macfarlane, Cummings, & Allison, 1986: Smith & Macfarlane, 1996).

Guarner and Malegelada (2003) also point out that most of the body's immune cells lie in the mucosa of the gastrointestinal tract, and it is here that communication between mucosa and microflora is vital for the immune system to be effective. However, it appears that this communication needs to be started early in life in order for the immune system to be constructed, and for it to develop appropriately (57 Sudo et al., 1997).

Communication

To maintain a competent digestive tract, communication between beneficial bacteria, and also between bacteria and the immune system within the digestive tract is essential. To obtain certain carbohydrates, *Bacteroides thetaiotamicron* signal the epithelial cells of the gut lining to take up the required carbohydrates (Falk, Hooper, Midvedt, & Gordon, 1998). To repel pathogenic invaders, resident bacteria can interfere with the invaders signaling devices (quorum sensing) so they have difficulty in knowing how many of the pathogens are present in order to take over territory. In the meantime, the resident bacteria compete effectively for nutrients, secrete toxins for pathogenic demise, and help protect the immune system embedded in the lining of the digestive tract (Neish, 2002).

As important players in the immune system, e.g. in cases of infection, monomorphic major histocompatability complex class 1 related (MR1) molecules reside in the intestinal lining of the gastrointestinal tract. MR1 molecules select or restrict mucosal-associated invariant T (MAIT) cell proliferation in the gastrointestinal tract. However, bacteria are necessary to provide expression on the surface of MRI molecules which results in MAIT production and proliferation, and promotes defensive action at the site where pathogenic bacteria attempt to penetrate the mucous membrane of the intestines. It was found in mice, that if no microflora were present, there were no MRI molecules expressed and therefore no MAIT cells produced. Conclusion MAIT cells rely on MR1 molecules and bacteria to respond to infection (Treiner, et al. 2003).

Cancer Risk and Prevention

Research on animals, reviewed by Guarner and Malagelada (2003), reveals that the presence of bacteroides and clostridium strains of bacteria apparently increase the risk of colon cancer, whereas the presence of strains such as lactobacillus and bifidobacteria appear to decrease the risk of tumor development, tumor growth, and reduced colon inflammation (Horie, et al., 1999; O'Mahoney, et al., 2001; Onou, Kado, Sakaitani, Uchida & Morotomi,

1997; Pool-Zobel, Neudecker, & Domizlaff, 1996; Singh, et al., 1997). Further the reviewers describe a study of humans (Moore & Moore, 1995) in which *Bacteroides vulgatus* and *Bacteroides stercoris* appeared to increase risk of cancer, while *Lactobacilllus acidiphilus* SO6 and *Eubacterium aerofaciens* appeared to decrease the risk of cancer. The conclusion of the review is that the presence or absence of certain bacteria may influence the risk of colon cancer. However, the studies of Van Leeuwen and Berg (as cited in Guarner & Malagalada, 2003) show that even healthy bacteria can become lethal if the lining of the gut is compromised, and bacteria are able to pass into the bloodstream (bacterial translocation) causing painful medical problems, even death.

It has been known for many years that pathogenic bacteria could be used to combat many diseases but have proved too toxic for use in humans. Now, because of the ability to penetrate the body's defense system, researchers are discovering new ways to introduce components of the pathogens to combat cancer and other diseases (Critchley, et al., 2004; Radford, et al., 2002).

E. coli, at times a virulent pathogen, when used as a vector, has the distinct advantage of carrying large recombinant protein molecules. However, the researchers posit that the *E. coli* vectors could also be used to transport naturally occurring proteins from bacteria (Critchley, et al., 2004). In other research, the pathogenic parts of *E. coli* were removed, and ovalbumin proteins and the lysing components of the pathogenic bacteria, *Listeria monocytogenes,* were enclosed in *E. coli* shells. The manipulated vectors were then inserted into mice already injected with ovalbumin tagged melanoma cancer cells. The vectors were almost immediately engulfed by macrophages which surrounded the vectors in membrane enclosures. The *Listeria monocytogenes* lysed the membranes, and freed the proteins. The proteins were then tagged as antigens, which resulted in the immune system killer cells searching out and destroying the ovalbumin coated melanoma cells. Mice injected with the vectors had no tumors, even after 90 days. The control group mice died within 16 days. The researchers suggested that other cancers and diseases could be controlled using this vector model with differing strains of bacteria (Radford, et al., 2002).

In other research, an energy producing protein, azurin, secreted by the bacteria, *Pseudomonus aerufinosa,* was put in *E. coli* shells which were then injected into mice that had been transplanted with melanomas. The azurin injections caused apoptosis in the cancerous cells. The tumors shrank and none of the mice died (Yamada, et al., 2002).

A vaccine for reducing tumors was developed by combining elements of two strains of pathogenic bacteria, *Listeria monocytogenes* and *Escheriscia coli,* and putting them into plasmid vectors. The vectors were then introduced into mice that had been infected with cervical cancer. After 63 days, the vaccinated mice were still alive. The control group were sacrificed because of growing tumors. The researchers posit that similar vaccines could be assembled using other bacterial strains (Versch, Pan, & Paterson, 2004). Guarner and Malagelada (2003) state that, to prevent diseases in the colon, it would be helpful to comprehend the interactions of bacteria and humans.

Bacteria and the epithelial lining of the digestive tract

Neish (2002), in a scientific review, points out that the interruption of helpful communication between the normally healthy bacteria and the epithelial cells that line the digestive tract can lead to disastrous results. The lining of the digestive tract consists of a single layer of epithelial cells, and its tight cell junction formation acts as a protective boundary between the digestive tract and the bloodstream. The other role of the epthielial cells is to allow fluids to enter or leave through this system. The epithelial lining also forms a barrier, similar to the blood brain barrier of the cranial brain, to protect the inner body from disease or pathogenic invasion (Hecht, 1999).

There is a mutually beneficial symbiotic relationship between eukaryote epithelial cells and enteric prokaryote microflora, which are in constant, immediate contact. Problems result if contact is compromised. The epithelial cells provide a mucus to which the bacteria can attach themselves. The microflora keep the surface of the epithelial cells clean by digesting the mucus. In return, the mucus helps to nourish the microflora. Some bacteria can signal the epithelial cells to take up carbohydrates which also provides nutrients for the microflora (Falk, Hooper, Midtvedt, & Gordon, 1998). Neish (2002) posits that the bacteria are directing the actions of the host to provide nourishment for themselves.

Wilson's studies (as cited in Neish, 2002) show how the microflora in the human digestive system help bile acids, bilirubin, and cholesterol to provide energy. The resident bacteria also assist in the synthesis of vitamin K which is necessary for the clotting of blood. The normal, healthy microflora of the colon aids in immunity against diseases. The resident bacteria within the enteric system prevent pathogenic bacteria from taking up residence on the epithelial cells of the gut lining by using up all available nutrients, blocking communication between the pathogens, or producing toxins for their demise. *Clostridium difficile* are not pathogenic in childhood, but in an adult can become pathogenic, mostly after the intake of antibiotics (Surawicz & McFarland, 1999).

The epithelial cells of animals bred without bacterial communities do not develop properly, but when microflora is introduced the epithelial cells develop normally (Falk, Hooper, Midvedt, & Gordon, 1998). Further, it was found that when bacteria *Bacteroides thetaiotamicron* was introduced into germ-free mice, they induced the epithelium cells to take up carbohydrates as nourishment for the bacteria. Also, the same bacteria, introduced into germ-free mice, affected the genetic mechanisms involved in taking in nourishment and building an effective epithelial barrier (Hooper, et al., 2001). As Neish (2002) points out, the resident microflora have the capabilities of manipulating the host cells.

Pathogenicity

In a weakened gut lining and immune system, as stated previously, the communication transport system is also employed to transmit pathogenic bacteria. After the Type III transport system (TTSS) penetrates the host cell, the inserted pathogenic proteins interrupt the host cell communication system, and set up their own lines of communication between the bacteria and the host organism (Viprey, Del Greco, Golinowski, Broughton, & Perret, 1998). At the same time the pathogenic proteins interfere with all the other cell activities (Hueck, 1998). After penetrating an organism, the invaders potentiate the release of a defensive inflammatory response by the organism, creating a wall around the infection to stop it spreading (Ciesiolka, et al., 1999). In a plant, this creates an enclosed niche where bacteria can be nourished by the plant's carbohydrates, and the plant is able to make use of the bacterial ability to fix nitrogen. Nitrogen is necessary for the survival of the plant, an example of an endosymbiotic relationship between plant and bacteria (Viprey, Del Greco, Golinowski, Broughton, & Perret, 1998). Pathogenic bacteria without TTSSs have been shown to be unable to establish symbiotic relationships with invaded organisms (Dale, Young, Haydon, & Wellburn, (2001). Therefore, besides it pathogenic abilities, it appears that TTSSs act as a communicator between prokaryotes and eukaryote cells (Neish, 2002).

Employing the TTSS communication system, the pathogenic proteins signal the activation or suppression of an inflammatory response in the epithelial cell lining for opposing reasons. Activation of the inflammatory response can deny entrance of pathogenic bacteria to the internal body structure by enclosing and destroying invading bacteria. However, if too much inflammation is caused pathological damage can ensue. The opposite action, suppression of the inflammatory response, can allow pathogenic bacteria to penetrate the epithelial barrier and spread infection (Ciesiolka, et al. 1999: Viprey, Del Greco, Golinowski, Broughton, & Peret, 1998).

Each pathogenic bacteria has the same TTSS method of delivery with similar virulence genes, even in distantly related bacteria. However, each strain of bacteria delivers their proteins in different configurations or combinations and this results in myriads of different diseases in both plants and animals (Hueck, 1998).

Researchers have identified a protein kinase enzyme that kills macrophages which respond first to the body's immune system. The role of the macrophages is to warn the immune system so that defensive cells are alerted. If pathogens attenuate the macrophage alarm system, the immune system is unaware of the bacterial invasion, and infection or disease ensues. Macrophages possess a Toll-Like Receptor 4 (TLR4) that can either signal life or death to themselves. Three strains of bacteria, that cause either anthrax, bubonic plague, or typhoid fever, can effectively activate the death signal and the macrophage becomes a victim of apoptosis. The activating instrument on the bacteria is a dsRNA response kinase (PKR). The researchers posit that a case of viral flu could be followed by a serious bacterial infection (Hsu, et al., 2004).

The lipopolysaccharide coated surface of invading pathogenic bacteria causes cytokine release by macrophages. The cytokine release signals the defending immune cells about the pathogenic invasion. However, scientists discovered that *Bacillus anthracis*, that causes anthrax, has the ability to cleave the macrophage protein MKK6 in the IRF3 pathway. MKK6 protein is one of the proteins needed to produce cytokines. With the alert system deactivated, there is no immune response, and therefore, pathogenic proliferation occurs in the circulatory system (Dang, David, Navarro, & Anderson, 2005).

Probiotics, Prebiotics, & Synbiotics

Live cultures of bacteria, known as probiotics, are being increasingly investigated by research and clinical trials to scientifically show if introducing missing or depleted non-pathogenic bacteria, or implementing levels of resident microflora in the digestive tract are a positive health benefit (Collins & Gibson, 1999). Probiotics are eaten in food products. Yogurt has been the traditional method of delivery, but now probiotics are introduced in fruit drinks, and also infant formulas (PDR Health, 2005).

Introduced as health promoting adjuncts to already resident bacteria (Guarner & Schaafsma, 1998), probiotics are now promoted as helpful in increasing nutrition and immune factors (PDR. 2005). Found to improve the treatment of diarrhea, probiotics have been administered in both young and old. Bacteria in yogurt assists in the digestion and absorption of lactose in lactose intolerant people (Guarner & Malagelada, 2003). However, if too many probiotics are ingested, there may be undesirable side effects that could lead to risk of infection (Guarner & Schaafsma, 1998).

Yogurt is commonly recommended to be taken after ingestion of prescribed antibiotics. However, probiotics can now be administered in capsule or tablet form, each product claiming to be more effective than the other. It is also suggested that probiotic products be date stamped because there may be a time factor of efficacy involved (PDR Health, 2005).

In humans, administration of probiotics improved post-operative recovery, reduced infections, resulted in shorter recovery times in pancreatitis, and subdued diarrhea induced by antibiotic treatment. Also, probiotics reduced infectious diarrhea. Success in treating pathogenic *Clostidium difficile*, was only successful after the second infection. Diarrhea that resulted from antibiotic or radiation treatment, was also successfully treated after probiotic ingestion. In animals, inflammation caused by inflammatory bowel disease was reduced. Probiotics were also successful in the treatment of *Helicobactor pylori* infection (Fedorack & Madson, 2004; Lippencott, Williams, & Wilkins, 2004).

The probiotic cultured products, while both in storage and use, must have cells that are capable of normal development in the digestive tract, and that also add to the beneficial actions of the resident microflora. Ingested probiotics must survive the acidity of the stomach, have the ability to attach to the digestive tract lining, inhibit or destroy pathogenic bacteria, and be able to assist in immune functions (Collins & Gibson, 1999). There is also much discussion over the benefits of mixing several strains of bacteria at the same time (PDR Health, 2005).

If probiotic ingestion is to be successful, the strain of the bacteria required must be known because each probiotic is strain specific. A new area of investigation is possible since the identification of 16S ribosomal RNA and its capabilities of identifying the capabilities of the different species of bacteria while they are actually functioning within the gastrointestinal tract. Problems had arisen when trying to identify bacterial capabilities in laboratory cultured specimens. Now it is possible to identify probiotic additions and its interactions with resident microflora (Collins & Gibson, 1999).The Toll like receptors of the epithelial and immune cells are able to recognize the differences between resident and introduced bacteria. Many resident bacteria can block pathogenic bacteria from attaching to the cells that line the digestive tract, or from penetrating them (Fedorack & Madson, 2004; Lippencott, Williams, & Wilkins, 2004).

Research reviewed by Guarner and Malagelada (2003) showed that toxic effects of faecal enzymes that promote tumors in humans were reduced by ingestion of probiotics (Goldin, & Gorbach, 1984; Marteau, Pochart, Flourie, Schumann, & Quignon, 1997). Concerning animals, research reviewed by U.S. Probiotics (2005) reveals that probiotics are effective in response to antibiotic intake, cases of diarrhea, irritable and inflammatory bowel syndrome, and also in cases of mild hypertension. Only one project was carried out on humans and it was found that re-occurrence of symptoms of cancer appeared after a longer period of time than in controls.

Prebiotics, which are non-digestible food products, stimulate the growth of resident colonic bacteria (Tannock, 1995), but, like probiotics, are strain specific (PDR Health, 2005). Prebiotics are composed of fructooligosaccharides or galactooligosaccharides. In animals and humans, successful administration of prebiotics, each to its correct strain, has shown some promotion of anti-carcinogenic activity, prevention of pathothegenic bacterial adherence to the gut lining, and reduction of triglyceride and cholesterol levels. They have also shown increased absorption of calcium and magnesium, increased immunity to pathogenic bacteria, and lowered adverse symptoms in some cases of Irritable Bowel Syndrome Prebiotics, administered for ulcerative colitis, showed improvement in epithelial cells, increased the amount of short chain fatty acids, and encouraged growth of probiotic bacteria. There were extremely few negative results in the research when prebiotics were administered (Fedorak& Madson, 2004; Lippencott, Williams, & Wilkins, 2004).

The administration of probiotics combined with same-strain prebiotics is called synbiotics. The aim of the synbiotics is to replace missing bacteria, to ensure their growth, to decrease pathogenic invasion, and to ensure proliferation of healthy bacteria. Probiotics combined with prebiotics results in a more positive treatment in several areas. Synbiotic treatment reduced toxic effects, repaired DNA damage, and lowered the proliferation of cancerous cells in colon polyps (Nutra USA, 2004). In cancer research on animals, synbiotics were found to reduce tumor growth, reduce enzyme activity that promoted cancer growth, lessen DNA damage in colon cells, and increase immune function (Van Loo, 2005).

Enteric Nervous System

An increasing number of scientists are now investigating the re-discovered role of the enteric nervous system in the human gastrointestinal tract. They are building on the research of forgotten scientists from over a century ago e.g. Bayliss and Starling, Trendelenberg, and the original editor of the Journal of Physiology, J. Langley.

Within the gastrointestinal tract, Goyal & Hirano (1996) posit, the enteric nervous system (ENS) "controls the motility, exocrine, and endocrine secretions, microcirculation, regulates immune and inflammatory process" (p. 1106). The extremely complicated ENS has major and minor pathways within the walls of the gastrointestinal tract that run from the esophagus to the anus. One major pathways is the myenteric plexus which, with its motor neurons, controls the downward movement of contents. The submucosa plexus, the other major pathway, aided by its sensory neurons, registers and recognizes the contents passing through the digestive tract. Another set of neurons, the interneurons, pass signals between the myenteric plexus and the submucosa plexus to ensure cooperation, and the smooth movement of the colon contents. The plexuses have the same glial cells as the cranial brain, the cells that nourish nuerons (Bowen, 2004).

The central nervous system (CNS) and the enteric nervous system (ENS) originate in the same embryonic segment of tissue, the neural crest, before migrating to their final destinations (Gershon, 1999). The digestive tract has evolved to interact with the outside world. Like the CNS, the ENS has a tightly jointed epithelial barrier that prevents access by harmful substances (Neish, 2002). The ENS has over 100 million neurons, many more than the spinal chord. It produces the same neurotransmitters as the brain, including serotonin, dopamine, glutamate, norepinephrin, and nitric oxide. The ENS also contains hormones, endorphins, and neuropeptides. Neuropeptides are brain proteins, the intermediaries in hunger, thirst, etc. In the ENS, there are also benzodiazapines; pain relievers that travel to the cranial brain benzodiazapine receptors. Antigens can also be created within the digestive tract to produce immunity from further virulent attacks (Blakeslee, 1996).

ENS neurons transmit messages from one section of the digestive tract to the other. Jackie Wood, of The Ohio State University, (as cited by Blakeslee, 1996) says that the ENS has a programmed set of simple patterns that remit specific instructions. The same simple patterns are found in lower evolved invertebrates. When extreme stress occurs, mast cells with the aid of histamines, call on the immune system. A response is orchestrated that sets up a barrier of inflammation that protects the bloodstream from dangerous infection. Psychoactive drugs that prevent the uptake of serotonin in the ENS, create a serotonin flood of stimulation that leads to diarrhea and/or nausea (Blakeslee, 1996).

The cranial brain and the enteric brain both have the same sleep patterns. After falling asleep, both systems slow down. At 90 minute intervals there are rapid eye movements showing increased activity in the cranial brain, and also at 90 minute intervals there are increased motility actions in the digestive tract. In his book, *Patient Heal Thyself,* (as cited in Angelmed, 2005) Rubin explains that in Indian and Chinese cultures, the abdomen is revered as the reservoir of healing and energy. Deities are portrayed with protruding stomachs because of their abundance of prana (India) and din tian (China). In China, it is believed that the din tian is where the roots of longevity and accrued wisdom grow.

The ENS, through its sensory systems, is capable of effecting its own survival actions. If the vagus nerve, the only connection to the cranial brain, is severed, the ENS can still carry out its appointed tasks. There is no other organ that can do this. If the nerves to the brain are cut off from any other organ in the body, the organ dies (Gershon, 1998). Hirschsprung's disease occurs where the part of the colon nearest to the anus has no nerves, and is incapable of a bowel movement. Jackie Wood (as cited in Whyfiles, 2005) tells how a German surgeon cut out the faulty section of the colon and reattached the healthy part of the colon to the anus. Just over a year later, normal bowel movements began. The neurons in the upper part of the colon had learned a new behavior.

In another set of events, as quoted in Blakeslee (1996), Dr. Gershon relates how, in a military hospital, paraplegics who had enemas to remove fecal compactions were put on a routine of having their enemas at 10 a. m. every morning. When the routine was to be changed to using enemas only when the impactions occurred, all the patients had natural bowel movements at 10 a.m. the next day. To remember such actions takes memory, and memory involves learning. Again, the ENS demonstrated it is capable of learning.

Blood Brain Barrier

Medical research is extremely interested in the invasion of the brain by pathogenic bacteria, and also how medicine can effectively be delivered to the brain to relieve disease, inflammation, and injury. The gelatinous brain is surrounded and protected from outside injury by cerebrospinal fluid, three layers of tissue, and the bony skull. A tightly knit band of brain microvascular or capillary endothelial cells, more commonly known as the blood brain barrier (BBB), protects the brain from pathogenic invasion, but allow nutrients such as glucose, oxygen, and iron to cross into the brain. Unfortunately, because they are composed of minute molecules, alcohol, nicotine, and various other addictive drugs are also able to cross. Large molecules and most harmful substances under normal circumstances cannot penetrate the blood brain barrier (Gershon, 1999).

There are no resident bacteria in the brain. However, the pathogenic bacteria *Group B Streptococcus* (GBS), which causes meningitis, can penetrate the blood brain barrier. When a large enough number of GBS have congregated, penetration ensues. Each GBS, is encapsulated in a vacuole that has a protein coating that mimics the host, which results in a lowered immune response, and so allows the GBS passage through the blood brain barrier (Pondrom, 2003).

Research at the University of Maryland at San Diego, found that the two proteins, zonulin and zot, can be manipulated to unlock the tightly bound cell junctions in the cranial blood brain barrier. The opening of the cell junctions would allow large molecules of medicine to pass through to the brain to help alleviate chronic brain diseases and injury. In a naturally occurring sequence, the proteins zonulin and zot, by attaching themselves to the tightly bound epithelial cell junctions in the digestive tract, enable nutrients to cross through the opened junctions into the blood stream (Levitt, 2000). Another example showing CNS and ENS similarities.

Bacteria are used in medical research that involves cell transportation. Recent research shows that when manipulated intestinal bacteria are added to stem cells, a large amount of neuropeptides and protein vehicles can be produced that are able to cross the blood brain barrier. The newly introduced neuropeptides can then initiate the healing process. Neuropeptides coordinate brain and body interactions. Composed of small pieces of protein, neuropeptides potentiate the healing process of brain cells damaged by disease, e.g. meningitis, Parkinson's disease, or Alzheimer's disease. Also, neuropeptides repair brain cells damaged by injury or prolonged use of illegal drugs. Unfortunately, neuropeptides cannot cross the BBB, and to obtain and administer them directly into the brain is expensive. Further research revealed that neuropeptides are capable of revitalizing sick and aging cells. Bacteria and stem cells were used to engineer neuropeptides and protein vehicles that crossed the blood brain barrier, and helped revitalize degenerating brain cells. It was postulated that if a greater range of neuropeptide/protein vehicles could be engineered with the use of bacteria and stem cells, many improvements could be made in the health and lifestyles of the human population (Natural Healing Research, April 29, 2005).

Chapter 4

Bacteria have the ability to survive independently, but they are also an indispensable adjunct to every other life form. All fungi, slime molds, plants, marine life, and animals, both invertebrate and vertebrate, require the presence of bacteria to survive and reproduce healthy offspring.

To replace the perception of bacteria as pathogenic and destructive, more education and enlightenment should be placed on the benefits that healthy bacteria provide. Probiotics, prebiotics, and synbiotics are helpful in repairing damaged intestinal flora. However, it would be more appropriate to take measures to keep the microflora in a healthy state from birth and throughout life. Also, more awareness should be made of the dangers that are a result of eliminating healthy bacteria.

To counter modern practices of eliminating bacteria as pathogens, mainly through the use of antibiotics, emphasis should be placed on protecting and nurturing healthy bacteria. Humans have evolved with bacteria as a vital adjunct. Ensuring successful propagation of healthy bacteria, through appropriate nutrition, would lead to less illness in the human population. For optimum health and well-being, more care should be devoted to the health and well-being of the bacteria that reside within the human body, with special emphasis on the digestive tract.

Medication is usually prescribed for emotional disturbance in order to control neurotransmitter output. Neurons, neurotransmitters, and neural receptors in the intestinal tract are the same as the ones in the cranial brain. It would therefore be advisable to consider the effect of medication on the digestive tract. Nutrient intake to improve mood should also be considered in psychiatric or psychological recommendations.

The highest number of bacteria in the human body reside in the intestinal tract, and because the digestive tract has its own enteric nervous system, it is imperative that education be carried out to make people aware of the initiatives that need to be taken to ensure proper nutritional intake for the maintenance, care, and overall health of the human population. To achieve a healthy lifestyle and to live in a world free from toxic wastes and pollution, it is imperative that everyone be informed, and understand, about the how, where, and why of the most important past and present members of our family, namely bacteria. They are vitally important to the mechanisms of our bodies Also, without bacteria the planet would be overwhelmed with detritus and garbage, and life for us would be impossible.

Due to its similarities to the cranial brain, and its independent neuronal system in the digestive tract, it is possible the enteric brain within the digestive system is the first brain. The cranial brain could have evolved later as the second brain in order to be an outpost. Residing in the most forward part of the body, the second brain would be in the optimum position to alert the organism to the presence of food or danger,

There are no resident bacteria in the cranial brain. This may have been a survival strategy of bacteria because of the bacterial ability to become pathogenic. Was the cranial brain developed as bacteria free in order to prevent the toxic abilities bacteria possess from being released inappropriately into the brain? However, if the organism is perceived as maybe no longer viable, disease can enter the brain via an excessive production of pathogenic bacteria. Also, the tight capillary junctions of the blood brain barrier can be breached with the aid of the two proteins, zolon and zot, the same proteins the enteric nervous system uses to open the tight junctions of the epithelial lining for nutrient absorption. Also, because of their ability to disguise themselves to achieve bacterial invasion, the bacteria in the gastrointestinal tract evidently come already equipped for invading the brain if necessary.

It appears that control is still retained by resident bacteria if penetration of the blood brain barrier is possible when injury or disease renders the organism redundant. The assumption that if the host dies, the bacteria dies is incorrect. After the death of the host, bacteria can be re-circulated into another role in another organism, be it protoctist, fungus, plant, or animal. One is reminded here of the belief in re-incarnation.

Bacteria have survived and multiplied successfully for billions of years. They greatly outnumber all other organisms on the planet Earth, but a bacterium is not credited with a 'nervous system' as science defines the working mechanism and intelligence of a life form. Communication takes intelligence. Bacteria communicate with each other through signal induction and quorum sensing which are needed for obtaining nutrients and energy, and also for avoiding danger. To be able to move toward nourishment and away from danger requires the ability to recognize both entities. Recognition requires memory, memory requires learning, and both need a base of operations - in humans it is called a brain. Therefore it should be in order to come to the conclusion that a bacterium has some form of brain? What form of intelligence does a bacterium have that we have not yet recognized? Any organism that can employ autopoiesis to produce an exact living copy of itself, as the bacteria do, should be credited with having an extremely high level of intelligence.

Which did come first - the CNS or the ENS? If a flatworm comes into contact with an obstruction it will turn away. but when the head is cut off, it will keep bumping against the obstruction. What part of the nervous system is being used to keep the worm in motion? Has the worm retained some of the nerve net of the hydra in its digestive tract that it is now using without the aid of its look-out post? If so, may this not support the idea that the enteric nervous system developed as the first brain, and that the cranial brain later evolved as a look-out post for nutrients or danger?

We also tend to overlook the fact that bacteria are necessary to all life forms and none can survive successfully without them. It takes intelligence to ensure that your evolving shelters only survive when you are present. It is possible that it is the bacteria that are the directors and architects of each endosymbiotic relationship. Further, the majority of bacteria reside in the digestive tract, and if the enteric nervous system is the original brain, it may follow that it is the bacteria within the digestive tract which are directing and controlling the evolution, health, and well-being of all organisms including humans.

Bacteria can signal activation or suppression of an inflammatory response using the Type III secretion system, the same pathway system that is involved in communication between the bacteria and the epithelial lining of the digestive tract. If there is an inappropriate, over abundant response, it can lead to pathological damage. Although we, as humans, can recognize and defend pathogenic invasion in order to survive, do the bacteria perceive the organism as defective if it cannot effectively defend itself, and, therefore, is in need of termination? Because of the membrane structure similarity between humans and bacteria, the host is able to, and does, cooperate with admitting pathogens. Did the bacteria actually evolve the host so that the host cooperates when the bacteria considers it not viable anymore?

It is a scientifically accepted fact, that all life, including humans, are evolved from prokaryote bacteria. Throughout the evolution of organisms that inhabit the planet Earth, bacteria have maintained control of each organism in its methods of replication, and in how its nutrients are obtained and absorbed. At the same time, bacteria were creating appropriate shelters so that each organism was protected, and if still considered viable, was able to survive in instances of danger and adversity. If every organism on the planet contains bacteria, and if every organism could not grow, thrive, or reproduce without their presence, and it is bacteria that can reduce the number of organisms in times of stress or overpopulation, it could be said that bacteria are in charge of life and death. They can create life and they can ultimately destroy life.

Apparently, the popular agreement between evolutionary scientists is that life is not evolving towards any particular goal (Margulis & Sagan, 1991). The argument is put forward that acquiring another set of genomes is not planned, but advantage is taken of the prevailing conditions. As the authors state, "Nature tends to be opportunistic not farsighted" (Margulis & Sagan , 2002, p. 42). Also, biologists cannot agree on whether sex has any evolutionary purpose. Margulis and Sagan (1991) argue that it does not. However, if bacteria have ensured their own survival for millions of years in a constantly changing environment by building ever more complicated shelters, until they have constructed a shelter that can effectively return the planet to an atmosphere of 1% oxygen, in which they can survive independently, it appears that they do have a determined plan of action.

Oxygen is lethal to archaebacteria, and if bacteria have created increasingly complicated shells or habitats to protect themselves in a world of increasing oxygen, have they finally created a human form that is now capable of producing the means of returning the atmosphere to less than 1% oxygen through pollution; an ever enlarging hole in the ozone layer; or a nuclear winter. Such an atmosphere would enable archaebacteria to return to their original independent existence. If this is the case, does it remove the human form from the role of host, and instead put bacteria in the role of architects and directors of the symbiotic relationship between bacteria and humans? Would this mean that it is the bacteria that are in control of the life, evolution, and eventually death of each and every organism on the planet Earth?

References

Abouheif, E. (2004). A framework for studying the evolution of gene networks underlying polyphenism: Insights from winged and wingless ants. In B. K. Hall, R. D. Pearson, & G. B. Muller (Eds.), *Environment, development, and evolution: Toward a synthesis* (pp. 123-137). London: MIT Press.

Akman, L., & Askoy, S. (2001). A novel application of gene arrays: *Escherichia coli* array provides insight into the biology of the obligate endosymbiont of tsetse flies. In S.E. Lindow (Ed.), *Proceedings of the National Academy of Sciences, USA. 98* (13) (pp. 7546-7551).

Angelmed, (2005). The brain gut connections. *Alternative medicine is God's medicine.* Retrieved from the World Wide Web, Sept 1, 2005. http//www.amc@angelmed.org

Arber, W. (2002). Evolution of prokaryote genomes. In R. W. Compans, M. Cooper, Y. Ito, S. Koprowski, F. Melchers, S. Olsnes, et al.(Series Eds.) & J. Hacker & J. B. Kaper (Vol. Eds.), *Current topics in microbiology and immunology: Vol. 1. Pathogenicity islands and the evolution of pathogenic microbes* (pp. 1-14). New York: Springer-Verlag.

Arthur, W. (2002). The emerging conceptual framework of evolutionary developmental biology. *Nature, Vol. 415,* Issue 6873, 757-764.

Asfaw, B., White, T., Lovejoy, O., Latimer, B., Simpson, S., & Suwa, G. (1999). *Australopithecus garli:* A new species of early hominid from Ethiopia. *Science, Vol. 284*, Issue 5414, 629-635.

Balon, E. K. (2004). Alternative ontogenies and evolution. In B. K. Hall, R. D. Pearson, & G. B. Muller (Eds.), *Environment, development, and evolution* (pp. 37 - 66). London: MIT Press.

Barghoorn, E. S. (1992). The antiquity of life. In L. Margulis & L. Olendzenski (Eds.), *Environmental evolution: Effects of the origin and evolution of life on planet earth* (pp.70 -84). Cambridge, MA, London: The MIT Press.

Bayliss, W. M., & Starling, E. H. (1899). The movements and enervation of the small intestine. *Journal of Physiology, Vol.* XX1V, 99-143.

Bengmark, S. (1998). Ecological control of the gastrointestinal tract: The role of probiotic flora. *Gut, Vol. 42*, No. 98, 2-7.

Blakeslee, S. (1996). 2nd brain in the stomach. *New York Times*. January 23, 1996. Retrieved from the World Wide Web, Sept. 1, 2005. http://www.aikido.com

Bowen, R. (2004). The Enteric Nervous System. Retrieved from the World Wide Web, August 21, 2005. http//www.physiology.org

Brace, G. L. (1988). *Stages of human evolution*. Englewood Cliffs, NJ: Prentiss-Hall.

Budi, S. W., van Tuinen, D., Martinotti, G., & Gianinazzi, S. (1999). Isolation from the *Sorghumbicolor* mycorrhizosphere of a bacterium compatible with arbuscular mycorrhiza development and antagonistic towards soil borne fungal pathogens, and stunts the growth of parasite P.parantila. *Applied Environmental Microbiology, 65*, (11), 5148-5150.

Carlsson, J. (2000) Growth and nutrition as ecological factors. In H. K. Kuratmizu & R .P. Ellen (Eds.) *Oral bacterial ecology: The molecular basis* (pp.67 - 130). Norfolk, England: Horizon Scientific Press.

Carneil, E. (2002). Plasmids and pathogenicity islands of *Yersinia*. In R. W. Compans, M. Cooper, Y. Ito, S. Koprowski, F. Melchers, M. Oldstone, et al. (Series Eds.) & J. Hacker & J. B. Kaper (Vol. Eds.), *Current topics in microbiology and immunology: Vol. 1. Pathogenicity islands and the evolution of pathogenic microbes* (pp. 1-14). New York: Springer-Verlag.

Cavanaugh, C. M. (1995). Harvard University. In D. Stover's Creatures of the thermal vents. *Popular Science, May 1, 1995*. Highbeam Library. http://www.gene@seawife.gsfc. nasa.gov Retrieved from the World Wide Web June 23, 05.

Chisholm, S. (2003). The cells that rule the seas. *Scientific American* Retrieved January 25, 2004 from the World Wide Web http://www.ScientificAmerican.com

Ciesiolka, L., Hwin, T., Gearlds, J., Minsavage, G., Saenz, R., Bravo, S., et al. (1999). Regulation of expression of a virulence gene avrRxv and identification of a family of host interaction factors by sequence analysis of avrBsT, *Molecular Microbe-Plant Interactions, 12*, 35-44.

Clement, S. (1996). Magnetic microbes. *The curious microbe*. Retrieved from the World Wide Web June 24, 2005. http://www.commtechlab.msu.edu

Collins, M. D., & Gibson, G. R. (1999). Probiotics, prebiotics, and synbiotics: Approaches for modulating the microbiology of the gut. *American Journal of Clinical Nutrition, Vol.69*, No. 5, 1052 -1057.

Cort. J. (Writer/Producer/Director). (2005). Car talk - fuel cells [Television series episode]. In Irwin, D. (Producer), & Robertson, M. (Co-Producer), *Nova science now*. Boston: WGBH. Educational Foundation.

Crews, D., Fleming, A., Willingham, E., Baldwin, R., & Skipper, J. K. (2001). Role of steroidogenic factor 1 and aromatase in temperature-dependent sex determination in the red-eared slider turtle. *Journal of Experimental. Zoology, 290,* 597-606._

Critchley, R. J., Jezzard, S., Radford, K. J., Goussard, S., Lemoine, N. R.., Grillot-Courvallin, S., et al. (2004). Potential therapeutic applications of recombinant, invasive *E. coli. Gene Therapy, Vol. 11* (15) 1224-1233.

Dale, C., Young, S. A., Haydon, D. T., & Welburn, S. C. (2001). The insect endosymbiont *Sodalis glossindius* utilizes a type III secretion system for cell invasion. *Proceedings of the National Academy of Sciences, USA, 98,* (pp. 1883-1888).

Dang, O., David, M., Navarro, L., & Anderson, K. (2005). Study by UCSD gives new insight into how anthrax bacteria can evade a host's immune response. Retrieved from the World Wide Web, Jan.28, 2005. http://www.biology.ucsd.edu.news/article_010604.html

Darling, K. (2000). *There's a zoo on you.* Brookfield, CT: The Millbrook Press.

Darwin, C. (1859). *The origin of species.* New York: Random House (1993).

Davis, P. W., & Solomon, E. P. (1986). *The world of biology.* (3rd ed.) New York: Saunders College Publishing.

Falk, P., Hooper, L., Midvedt, T., & Gordon, J. (1998). Creating and maintaining the gastrointestinal ecosystem: What we know and need to know from gnotobiology. *Microbiology and Molecular Biology Reviews, 62,* Issue 4, 1157-1170.

Fedorack, R., & Madsen, K. (2004). Naturally occurring and experimental models of inflammatory bowel disease. In J. Kirsner (Ed.), *Inflammatory bowel disease* (pp. 113-143). Philadelphia: W. B. Saunders.

Feferman, L., & Stobie, J. (Producers). Ferris, T. (Creator/Host). (1999). *Life beyond earth.* [Video]. Distributed by Warner Home Video, 4000, Warner Blvd., Burbank, CA, 91522.

Fortey, R. (1997). *Life. A natural history of the first four billion years of life on earth.* New York: Alfred A. Knopf.

Fukatsu, T., & Nicoh, H. (2000). Endosymbiotic microbiota of the bamboo Pseudococcid *Anotnina crawii,* (insecta, Homoptera). *Applied and Environmental Microbiology, Vol.66,* No. 2, 643-650.

Gershon, M. D. (1998). *The second brain.* New York: HarperCollins.

Goldin, B. R., & Gorbach, S. L. (1984). The effect of milk and *Lactobacillus* feeding on human intestinal bacterial enzyme activity. *American Journal of Clinical Nutrition, Vol. 39,* 756-761.

Goyal, R. K., & Hirano, I. (1996). Mechanisms of disease. *New England Journal of Medicine. Vol. 334,* 1106-111.

Graham, J. B., Dudley, R., Aguilar, N. M., & Gans, C. (1995). Implications of the late paleozoic oxygen pulse for physiology and evolution. *Nature, 375,* 117-120.

Groisman, E. A., & Ochman, H. (1996) Pathogenicity islands: bacterial evolution in quantum leaps. *Cell, Vol. 87,* Issue, 5, 791-794.

Guarner, F., & Schaafsma, G. J. (1998). Probiotics. *International Journal of Food Microbiology, 39,* Issue 3, 237-238.

Guarner, F., & Malagelada, J. R. (2003). Gut flora in health and disease. *Lancet, Vol. 361,* Issue 9356, 512-519.

Hall, B. K. (2004). Introduction: Evolution as the control of development by ecology. In B. K. Hall, R. D. Pearson, & G. B. Muller (Eds.), *Environment, development, and evolution: Toward a synthesis* (pp. ix -xxiii). London: MIT Press.

Hargreaves, W. R., & Deamer, D. W. (1978). Origin and early evolution of bilayer membranes. In D. W. Deamer (Ed.), *Light transducing membranes: Structure, function, and evolution* (pp.23 -59). New York: Academic Press.

Harper, A. (Producer & Director), & McMaster, J. (Writer). (2005). *Origins: How life began* [Nova television series]. Boston: WGBH.

Heavey, P. M., & Rowland, I. R. (1999). The gut microflora of the developing infant: microbiology and metabolism. *Microbial Ecology in Health and Disease, 11,* 75-83.

Hecht, G. (1999). Innate mechanisms of epithelial host defense: Spotlight on intestine. *American Journal of Physiology, 277,* C351-C358.

Hooper, L. V., Wong, M. H., Thelin, A., Hansson, L., Falk, P. G., & Gordon, J. R. (2001). Molecular analysis of commensal host-microbial relationships in the intestine. *Science, 291,* 881-884.

Horie, H., Kanazawa K., Okada, M., Narushima, S., Itoh, K., & Terada, A. (1999). Effects of intestinal bacteria on the development of colonic neoplasm: An experimental study. *European Journal of Cancer Prevention, 8,* 237 -245.

Hrywana, Y. (1996). Giant bacteria inhabit fish guts. *The curious microbe.* Retrieved from the World Wide Web June 24, 2005. http://www.commtechlab.msu.edu

Hsu, L.C., Karin, M., Park, J. M., Zang, K., Luo, J. L., Maeda, S., et al. (2004). The protein kinase PKR is required for macrophage apoptosis after activation of Toll-like receptors. *Nature, Vol. 428,* Issue 6980, 341-345.

Huddler, G. W. (1998). *Magical mushrooms, mischievous molds.* Princeton, NJ: Princeton Press.

Hueck, C. J. (1998). Type 111 protein secretion systems in bacterial pathogens of animals and plants. *Microbial Molecular Biology Revue. 62*(2) 370-433.

Hurst, G. G. D., & Jiggins, F. M. (2004). Male-killing bacteria in insects: Mechanisms, incidence, and implication. *Journal of Emerging Infectious Diseases. C. D. C. Vol. 6.* No. 4, 461-468.

Ingersoll, M., Groisman, E.A., & Zychlinski, A. (2002). Pathogenicity islands in *Shigella.* In R. W. Compans, M. Cooper, Y. Ito, S. Koprowski, F. Melchers, M. Oldstone, et al. (Series Eds.) & J. Hacker & J. B. Kaper (Vol. Eds.), *Current topics in microbiology and immunology: Vol. 1. Pathogenicity islands and the evolution of pathogenic microbes* (pp. 1-14). New York: Springer-Verlag.

Julien, D. (2000). On mycorrhiza and roses. *Northwest rosarian.* Northwest Pacific District of the American Rose Society.18402, 100th Ave. Snowhomish, WA 98296-8038.

Kimball, J. (2004).The origin of life: Abiotic synthesis of organic molecules. Retrieved January 25, 2006 from the World Wide Web: http://users.rcn.com/jkimball. ma.ultranet/Biology Pages/A/AbioticSynthesis.html

Kingsley, R. A., & Baumler, A. J. (2002). Pathogenicity island and host adaptation of *Salmonella serovars.* In R. W. Compans, M. Cooper, Y. Ito, H. Koprowski, F.Melchers, M. Oldstone, et al. (Series Eds.) & J. Hacker & J. B. Kaper. (Vol. Eds.). *Current topics in microbiology and immunology: Vol. 1: Pathogenicity islands and the evolution of pathogenic microbes*: (pp. 67-87). New York: Springer-Verlag.

Koneman, E. W. (2002). *Other end of the microscope: The bacteria tell their own story.* Washington D. C: ASM Press.

Kuramitsu, H. K., & Ellen, R. P. (Eds.). (2000). *Oral bacterial ecology: The molecular basis.* Norfolk, England: Horizon Scientific Press.

Larsen, E.W. (2004). A view of phenotypic plasticity from molecules to morphogenesis. In B. K. Hall, R. D. Pearson, & G. B. Muller (Eds.) *Environment, development, and evolution: Toward a synthesis* (pp. 117 - 124).London: MIT Press.

Leadbetter, J. (1996). The curious microbe. *Deinococcus radiodurens.* Retrieved from the World Wide Web June 24, 2005. http://www.commtechlab.msu.edu

Levitt, E. B. (2000). *University of Maryland researchers discover "key" to blood brain barrier.* University of Maryland Press Release. January 3, 2000. Retrieved from the World Wide Web, January, 29, 2005.
http://www.umm.edu/lnews/release/bloodbrain.html

Linderman, R. (2005). The mycorrhizasphere phenomenon. *Summary from a recent lecture in Mexico.* Personal correspondence received July, 2005. Inquiries to Robert Linderman, PhD. USDA ARS Horticultural Crops Research Laboratory, 3420 NW Orchard Ave., Cornwallis, Oregon, USA 97330.

Lippincott, W., Williams, T., & Wilkens, J. (20004). Mechanisms of action of probiotic bacteria. *Current Opinions Gastroenterology, 20*(2), 146-155.

Macfarlane, G. T., Cummings, J. H., & Allison, C. (1986). Protein degradation by human intestinal bacteria. *Journal of General Microbiology, 132,* 1647-1656.

Mamatha, G., Bagyaraj, D. J., & Jaganath., S. (2002). Inoculation of field established mulberry and papaya with arbuscular mycorrhizal fungi and a mycorrhiza helper bacterium. *Mycorrizha, Vol. 12.* No. 6, 313 -316.

Margulis, L. (1992). *Diversity of life: The five kingdoms.* Aldershot, UK: Enslow.

Margulis, L. (1992). Symbiosis theory: Cells as microbial communities. In L. Margulis& L. Olendzenski (Eds.), *Environmental evolution: Effects of the origin and evolution of life on planet earth* (pp. 149-172). Cambridge, MA: MIT Press.

Margulis, L., & Sagan, D. (1991). *Mystery dance: On the evolution of human sexuality.* New York: Simon & Schuster.

Margulis, L., & Sagan, D. (1995). *What is life.* New York: Simon & Schuster.

Margulis, L., & Sagan, D. (2002). *Acquiring genomes: A theory of the origins of species.* New York: Basic Books.

Margulis, L., & Schwartz, K. V. (1998). *Five kingdoms: An illustrated guide to the phyla of life on earth* (2nd ed.). New York: Freeman.

Marsh, P. D. (2000). Oral ecology and its impact on oral microbial diversity. In H. K. Kuramitsu & R. P. Ellen (Eds.), *Oral bacterial ecology: The molecular basis*(pp. 11-65). Norfolk, England: Horizon Scientific Press.

Marsh, A., Mullineux, L., Young, C. M., & Manahan, D. T., (2001). Larval dispersal potential of the tube worm Riftia pachyptila at deep sea hydrothermal vents. *Nature, 411,* 77-80.

Marteau, P., Pochart, P., Flourie, B., Schuman, S., & Quignon, V. (1997). Effect of chronic ingestion of a fermented dairy product containing *Lactobacillus acidophilus* and *Bifidobacterium bifidum* on metabolic activities of the colonic flora in humans.. *American Journal of Clinical Nutrition. 50*, 269-273.

Mayr, E.(2001). *What evolution is.* New York: Basic Books.

Monastersky, R. (1997). Deep dwellers: Microbes thrive far below ground. Retrieved from the World Wide Web January 24, 2006. http://www.sciencenews.org/pages/sn_arc97/3_29_97/bob1.htm

Moore, W. E., & Moore, L. H. (1995). Intestinal floras of populations that have a high risk of colon cancer. *Applied Environmental Microbiology, 61*, 3202-3207.

Natural Healing Research (2004). *Stem cell theory or neuropeptides.* Retrieved from the World Wide Web July 15, 2005. http://www.therapon1.com/stemcell.html

Neish, A. S. (2002). The gut microflora and intestinal epithelial cells: A continuing dialogue. *Microbes and Infection, Vol. 4,* Issue 3, 309-317.

Normak B. (2004). The strange case of the armored scale insect and its bacteriome. *PLoS Biology,* 2(3) e43. Retrieved from the World Wide Web Sept. 30, 2005. http://www.plosbiology.org/plosonline/?request=get-document&doi=10.1371/journal.pbio.002

Nutra USA (2004). Synbiotics could reduce risk of colon cancer. Retrieved from the World Wide Web August 15, 2005. www.nutraingredients-usa.com

O'Mahony, L., Feeney, M., O'Halloran, S., Murphy, L., Kiely, B., Fitzgibbon, J., et al. (2001). Probiotic impact on microbial flora, inflammation and tumour development in IL-10 knockout mice. *Alimentary Pharmacological Therapeutics, 15,* Issue 8, 1219-1225.

Onoue M., Kado, S., Sakaitani, Y., Uchida, K., & Morotomi, M. (1997). Specific species of intestinal bacteria influence the induction of aberrant crypt foci by 1,2-dimethylhydrazine in rats. *Cancer Letters , 113,* Issue 1, 179-186.

Oparin, A. I. (1953). *Origin of life.* New York: Dover Publications.

PDR Health (2005). Probiotics, prebiotics, and synbiotics: Harnessing enormous potential. Retrieved from the World Wide Web August 12, 2004.

Pollock, R. (1994). *Signs of Life.* New York: Houghton Mifflin.

Pondrom, S. (2003). UCSD researchers decipher function of blood brain barrier in bacterial meningitis. University of Maryland at San Diego Press Release, Sept. 2, 03. Retrieved from the World Wide Web, January 29,2006. *http://www.ucsd.news.edu/newsrel/health/Nizetweb/htm*

Ponnamperuma, C. (1992). Cosmochemical evolution and the origins of life. In L. Margulis & L. Olendenski (Eds.), *Environmental Evolution: Effects of the origin and evolution of life on earth* (pp.17 - 27). Cambridge, MA: MIT Press.

Pool-Zobel, B. L., Neudecker, C., & Domizlaff, I.(1996). Lactobacillus- and bifidobacterium-mediated antigenotoxicity in the colon of rats. *Nutrition and Cancer, 26,* 365-380.

Postgate, J. (1992). *Microbes and man (*3rd ed.*).* New York: Cambridge University Press.

Radford, R. J., Higgins, D. L., Passquini, S., Cheadle, E. J., Carta, L., Jackson, A. M., et al. (2002). A recombinant *E.coli* vaccine to promote MHC class 1 dependent antigen presentation: Application to cancer immunotherapy. *Gene Therapy, Vol. 9* (21), 1455-1463.

Reid, R. G. B. (2004). Epigenetics and environment: the historical matrix of Matsuda's pan-environmentalism. In B .K. Hall, R. D Pearson, & G. B. Muller (Eds.). *Environment, development, and evolution* (pp.7-35). London: MIT Press.

Roach, J. (2005). Hot water worms may use bacteria as shield. Popular Science. January 2005.Retrieved from the World Wide Web, January, 29, 06. http://news.nationalgeographic.com/news/2005/01/047_tueworms.html

Sankaran, N. (2000). *Microbes and people: An A-Z of microorganisms in our lives.* Phoenix, Arizona: Oryx Press.

Sapolsky, R. (2003). Bugs in the brain. *Scientific American,* March 2003, p. 97.

Sea & Sky (2005). Monsters of the deep: Giant tube worms. Retrieved from the World Wide Web June 24, 2005. http://www.seasky.org

Segelken, R. (1998). Acid relief for 0157:H7. Simple change in cattle diets could cut E.coli infection. USDA & Cornell Report. Retrieved from Worldwide Web. June 18, 2005. http://www.organicconsumers.org/organic/ecolimyths.cfm#Acid

Shackleton, N. J. (1995). New data on the evolution of Pliocene climatic variability. In E. S. Vrba, G. H. Denton, T. C. Partridge, and L. H. Buckle (Eds.), *Paleoclimate and evolution with emphasis on human origins (*pp. 242-248). New Haven, CT: Yale University Press.

Simon, G. L., & Gorbach, S.L. (1984). Intestinal flora in health and disease. *Gastroenterology, 86*: 174-193.

Singh, J., Rivenson, A., Tomita, M., Shimamura, S., Ishibashi, N., & Reddy, B.S. (1997). Bifidobacterium longum, a lactic acid-producing intestinal bacterium inhibits colon cancer and modulates the intermediate biomarkers of colon carcinogenesis. *Carcinogenesis, 18,* 833-841.

Singleton, P. (1999). *Bacteria in biology, biotechnology and medicine* (5th ed.). New York: Wiley.

Smith, E. A., & Macfarlane, G. T. (1996). Enumeration of human colonic bacteria producing phenolic and indolic compounds: Effects of pH, carbohydrate availability, and retention time on dissimilatory aromatic amino acid metabolism. *Journal of Applied Bacteriology, 81*: 288-302.

Snedden, R. (2000). *A world of microorganisms.* Hong Kong, China: Reed Publishing.

Snedden, R. (2003). *Cell division & genetics.* Chicago: Reed Publishing.

Stark, P. L., & Lee, A. (1982). The microbial ecology of the large bowel of breast fed and formula fed infants during the first year of life. *Journal of Medical Microbiology, Vol. 15,* 189 -195.

Stephenson, S. L., & Stempen, H. (1994). *Myxomycetes. A handbook of slime molds.* Portland, OR: Timber Press.

Stoppenbeck, T. S., Hooper, L. V., & Gordon, J. I. (2004). Developmental regulation of intestinal angiogenesis by indigenous microbes via Paneth cells. Retrieved from the World Wide Web August 12, 2005. Washington University In St. Louis, School of Medicine. http://www.reckessg@msnotes.wush.edu.

Stover, D.(2005). Creatures of the thermal vents. *Popular Science* (May 1, 1995). Highbeam Reasearch Library. Retrieved from the World Wide Web, 7/17/2006. http:/www.gene@seawife.gsfc.nasa.gov

Sudo, N., Sawamura, S., Tanaka, K., Aiba, Y., Kubo, C., & Koga, Y. (1997). The requirement of intestinal bacterial flora for the development of an IgE production system fully susceptible to oral tolerance induction. *Journal of Immunology, 159,* 1739-1745.

Surawicz, C., & McFarland, L. V. (1999). Pseudomembranous colitis: Causes and cures. *Digestion, 60,* Issue 2, 91-100.

Tannock, G. W. (1995). *Normal microflora: An introduction to microbes inhabiting the human body.* London: Chapman & Hall.

Torres, A. G., & Kaper, J. B. (2002). Pathogenicity islands of intestinal *E. coli*. In R. W. Compans, M. Cooper, Y. Ito, S. Koprowski, F. Melchers, M. Oldstone, et al. (Series Eds.) & J. Hacker & J. B. Kaper (Vol. Eds.). *Current topics in microbiology and immunology: Vol. 1. Pathogenicity islands and the evolution of pathogenic microbes* (pp. 1-14). New York: Springer-Verlag.

Treiner, E., Dubun, L., Bahram, S., Radosavljevic, M., Wanner, V., Tilloy, F., et al. (2003). Selection of evolutionarily conserved mucosal-associated invariant T cells by MRI. *Nature, 422*, Issue 6928, 164-169.

U. S. Probiotics (2005) Retrieved from the World Wide Web January 29, 2006 http://www.usprobotics.org

Van Loo, J. (2005). Synbiotics and cancer prevention in humans. European Union Syncan Project. Retrieved from the World Wide Web, August 14, 2005. e-mail jan.van. loo@orafti.com

Versch, T., Pan, Z., & Paterson Y. (2004). *Listeria monocytogenes* - based antibiotic resistance gene-free antigen delivery system applicable to other bacterial vectors and DNA vaccines. *Infection and Immunity, Vol. 72*, No. 11, 6418-6425.

Viprey, V., Del Greco, A., Golinowski, W., Broughton, W. J., & Perret, X. (1998). Symbiotic implications of type III protein secretion machinery in Rhizobium. *Molecular Microbiology, Vol. 28,* Issue, 6, 1381-1389.

Von Wintzingerode, W., Gerlach, G., Schneider, B., & Gross, R. (2002). Phylogenetic relationships and virulence evolution in the genus *Bordetella*. In R. W. Compans, M. Cooper, Y. Ito, S. Koprowski, F. Melchers, S. Olsnes, M. Oldstone, M. Potter, P. K. Vogt, H. Wagner (Series Eds.) & J. Hacker & J. B. Kaper (Vol. Eds.), *Current topics in microbiology and immunology: Vol. 1. Pathogenicity islands and the evolution of pathogenic microbes* (pp. 1-14). New York: Springer-Verlag.

Vrba, E. S. (1995a). The fossil record of African antelope (Mammalia, Bovidae) in relation to human evolution and paleoclimate. In E. S. Vrba, G. H. Denton, T. C. Partridge, & L. H. Buckle (Eds.), *Paleoclimate & evolution with emphasis on human origins (*pp. 385-424). New Haven, CT: Yale University Press.

Vrba, E. S. (1995b). On the connection between paleoclimate and evolution. In E. S. Vrba, G. H. Denton, T. C. Partridge, & L. H. Buckle (Eds.), *Paleoclimate & evolution with emphasis on human origins (*pp. 24-45).New Haven, CT: Yale University Press

Vrba, E .S. (2004). Ecology, development, and evolution: Perspectives from the fossil record. In B. K. Hall, R. D., Pearson, & G. B. Muller (Eds.) *Environment, development, and evolution; Toward a synthesis* (pp.85-105*).* Cambridge, MA: MIT Press.

Wassener, T. (2005). Ask a scientist archive. Retrieved from the World Wide Web, July 24, 2005. http://www.bacteriamuseum.org

Weinberg, S. (2000). *A fish caught in time.* New York: HarperCollins.

Whitney. E. N., Hamilton, E. M. N., & Rolfes, S. R. (1990). *Understanding nutrition* (5th ed.). St. Paul, MN: West Publishing.

Whyfiles. (2005). A brain in the gut. *Things that go bump in the night.* Retrieved from the World Wide Web, Sept, 1, 2005. http:/www.whyfiles.org/026fear/physiol.html

Wikipedia (2005). Genome. Retrieved from the World Wide Web January 29,06. http://en.wikipedia.org/wiki/Plasmid

Yamada, T., Goto, M., Punj, V., Zaborina, O., Chen, M. L., Kimbara, K., et al. (2002). Bacterial redox protein azurin, tumor suppressor protein p53, and regression of cancer. *Proceedings of the National Academy of Sciences, U.S.A. Vol. 99*, No. 22, 14098-14103.

Youson, J. H., (2004). Evolution of fish metamorphosis. In B. K. Hall, R.D. Pearson, & G.B. Muller (Eds.) *Environment, development, and evolution: Toward a synthesis* (pp. 239 - 277). Cambridge, MA: MIT Press.

About the Author

Dr. Abbott was born in Liverpool, England but now resides in the U. S. A. She has lived in several states and also in Iran. She holds a B. A. in Psychology, an M. Sc. in Counseling, and a Ph. D. in Nutrition. She counsels in weight loss, exercising, and stress management, combined with instructions in Hatha Yoga for balance in mind and body through breathing and stretching exercises. Her interests are golf, duplicate bridge, hiking, and travelling. Her book, The Nosmo King, a children's story to discourage smoking and the use of drugs, was published in 2001.